What a wonderful book *Wanderland* is! A witty, gentle, original and very modern quest for the magical (not the mythical) in Britain's landscape, which both made me laugh and moved me. I wish Roger Deakin could have read this book, for he would surely have recognised a kindred spirit in Jini Reddy.

Robert Macfarlane

An honest, contradictory and refreshing take on nature writing.

Noo Saro-Wiwa, *Conde Nast Traveller*

Witty and engaging.

Tom Robbins, *Financial Times*

She rejects the stereotypes placed on people of colour, and crafts a beautiful story of self-discovery and exploration of the natural world.

Anamika Talwaria, *Brown Girl Magazine*

*Wanderland* is extraordinary, unique even, standing apart from recent books about the British countryside… She is, she declares, lovesick: at times her prose has a dreamy, almost erotic charge.

Ben Hoare, *Countryfile*

Funny and touching.

Kathryn Hughes, *The Mail on Sunday*

Warm, open-minded and endlessly curious, Jini is an ideal guide to Britain's more unusual places and people. *Wanderland* is a truly engaging exploration, full of heart and soul.

Melissa Harrison

A candid, soulful and uplifting search for natural magic.

Stephanie Cross, *The Lady*

Curious and tenacious, Jini learns to accommodate both solitude and the gifts of chance, discovering at last a new way of being, a new way of seeing, a new way of listening to the complex voices of this archipelago – animal, aerial, human and other-than-human.

Katharine Norbury

A joyous celebration of the beauty we can see and the magic we can't.

Tay Aziz, *BBC Wildlife*

Jini has a wry, unique perspective on the nature of wilderness and the beauty of our landscape.

Sarah Barrell, *National Geographic*

A page turner.

Kirsten Jones, *Sunday Express 'S' Magazine*

With an unusual but timely eco-spiritual edge, and an alluring blend of memoir and nature-writing this touches on themes of identity and belonging as it charts how a restless spirit fell in love with her native land.

Caroline Sanderson, Editor at *The Bookseller*

*Wanderla* … so your mind to the p…

…er, *Wanderlust*

## A NOTE ON THE AUTHOR

**Jini Reddy** is an award-winning author and journalist. She was born in London to Indian parents from South Africa, and was raised in Montreal, Quebec, and now makes her home in the UK. Jini has a B.A. in Geography, an M.A. in English Literature and, increasingly, a passion for writing cross-genre, non-fiction narratives relating to landscape, travel, spirituality and culture. Her byline has appeared in the *Guardian*, *TIME*, *The Times*, *The Sunday Times Style*, *The Sunday & Daily Telegraph*, *Financial Times*, *National Geographic Traveller*, *Resurgence & The Ecologist* and many other publications. Her first book, *Wild Times*, was published in 2016 and she is a contributor to the *Winter: An Anthology for the Changing Seasons* and *Women on Nature*.

# WANDERLAND

## *A Search for Magic in the Landscape*

*Jini Reddy*

BLOOMSBURY WILDLIFE
LONDON • OXFORD • NEW YORK • NEW DELHI • SYDNEY

BLOOMSBURY WILDLIFE
Bloomsbury Publishing Plc
50 Bedford Square, London, WC1B 3DP, UK
29 Earlsfort Terrace, Dublin 2, Ireland

BLOOMSBURY, BLOOMSBURY WILDLIFE and the Diana logo are trademarks of Bloomsbury Publishing Plc

First published in Great Britain 2020. Paperback edition 2021.



A catalogue record for this book is available from the British Library

ISBN: PB: 978-1-4729-5195-3; HB: 978-1-4729-5193-9;
eBook: 978-1-4729-5194-6

2 4 6 8 10 9 7 5 3 1

Typeset in Bembo Std by Deanta Global Publishing Services, Chennai, India
Printed and bound in Great Britain by CPI Group (UK) Ltd, Croydon CR0 4YY

*In loving memory of Mandy Thatcher,*
*my cherished friend.*
*May your spirit soar.*

# Contents

# What Happens Up the Mountain Doesn't Always Stay Up the Mountain

A few years ago, I went up a mountain in the Pyrenees with a tent, nine bottles of water and almost no food. I wasn't being naïve or irresponsible, I simply wanted to commune with the wild in the raw.

It's a custom that has become quite fashionable these days in certain circles, even though it is as old as the hills. It wasn't the first time I'd done something like this, so I welcomed the experience. I kind of had an idea of what I was in for, in the way that if you've ever fallen madly in love, you know what it will feel like, even though every time it is completely different. This kind of experience wasn't about challenging myself – no, it was about quietening down, going inward and listening. No special skill required, which is just as well because I didn't have any, other than the ability to enjoy my own company. It didn't feel strange or alarming to be spending four nights on the mountain with only two apples, a handful of nuts, no phone and no watch, and those nine bottles of water.

Anyway, I needed the time out. I had a lot to get off my chest and I figured in the mountains I could cry my heart out. Up there alone on my first night, though, after the sun had gone down I heard a strange sound. It made my heart pound in a way that was nearly as frightening as the sound itself. That unearthly whisper on the other side of the canvas – well, my brain couldn't make sense of it. The guide who'd walked me up here had called the mountain 'Hartza Mendi', or 'Bear Mountain' in English. He'd spoken of the Lord of the Forest, a strange creature, the lovechild of Basque myth and the Pyrenean wilds. But I hadn't actually expected to hear its voice, if that's what 'it' was. It had come out of the dark, from nowhere. It was urgent and somehow… sentient. It was punctuated by pregnant silences that made me hold my breath as a wave of fear flooded my body. What do you do when you're in a blind panic? Me, I reached for a charm that was stashed in the tent pocket and I began to rock back and forth. Under my breath I muttered in a small, scared voice: 'I come in peace!' For once I was far too frightened to feel silly or self-conscious, my usual default setting.

Outside my tiny tent – weird, discombobulated voice aside – the mountain fell silent. No more gusts of wind, and whatever night creatures lived here and in the thick, now menacing, woods beside me were holding their breath. I'd heard no footsteps, no crackling of bushes, and anyway I'd been rooted to this spot on the flat top of this peak, like a landing strip for an alien craft, high in the mountains since noon. A mare and her foal had trotted up earlier to check me out or welcome me or show their concern for this strange woman 'stranded' in their territory, I wasn't sure which, but the only sign of human life I'd detected until now was the tinkle of a shepherd's bell in the valley down past the waterfalls and the emerald forest I'd walked through to get here.

A long minute or two after it began the voice stopped. Just like that. The mountain exhaled, the night sounds – noticeable only in their absence – started up again. Over the next four days and nights up here, I thought about my strange encounter and tried to make sense of it. Had some presence that made no sense to the rational side of my brain given me exactly what I'd hoped and prayed for before I walked up that mountain? I'd wanted – I'd *yearned* with my whole being – to hear nature's voice. Is that what I'd heard? Was it some kind of spirit? The Lord of the Forest? Who knows?

I've told this story to people and I know what most of them think: 'It was a bird, obviously. Or your imagination, silly New-Age deluded hippie.' Only, I could swear it wasn't. My proof? None, I had none. Only a deep conviction that what I heard that night wasn't a human or animal or bird but something quite mysterious and spirit-like. At any rate, after that experience, it was hard for me to just go for a walk or look at a tree or stare at the sky without hoping for an epiphany or some transcendent experience that would give me the feeling that the land was speaking to me in a way that went beyond the ordinary. I wanted to invoke *something* – for some life force to make its presence known to me – and the wanting of it felt like a kind of lovesickness.

Had I been a die-hard conservationist or scientist or maybe grown up on a farm, I'd have likely laughed myself silly at such notions. But those things hadn't been a part of my life. Instead, what I'd had was Hinduism and atheism by osmosis and then ordinary-growing-up secularism but with a yen for magical things. Call me sentimental but I wanted something more than to walk through an alluring landscape and admire its beauty. I wanted somehow to be more *porous*. I didn't want to be burdened by needing to know the name of every bird, creature, tree and petal. No, I wanted something else, something a bit Other and a bit mystical even – the seeking of it was what truly excited me.

As a travel writer, I had had experiences that opened my eyes and I was infinitely grateful for the world that revealed itself to me. Gul, my young blue-eyed hostess from the Kalash tribe deep in the remote valleys in Pakistan's North-West Frontier Province, had shared with me the ways of her people, their reverence for the gods of the river and the sun, and for their spirit ancestors. In Cape York in Queensland, I'd met two sisters who'd led me to a waterfall and told me a dreamtime creation story involving supernatural beings: they'd brought the earth's physical features into being, the sisters said. These encounters, and others, showed me as plain as day that for many indigenous people around the world all of nature was alive, imbued with spirit and a powerful ally if treated with respect. To some people I knew closer to home this idea made perfect sense. No big deal at all, but an obvious thing. But to many this was absurd. I never got why the words of people who live close to the land and treat her like kin – people who nurture an inner relationship with the earth – were rarely listened to or heeded beyond alternative circles. It's not like we couldn't use the input.

Still, despite those encounters abroad, back in the UK and sensitive to the mood of the day and the things I'd read and the voices I heard, I worried that I didn't love nature in the right way, that I didn't bring my gaze to bear upon Her in the *approved* way. What made me feel even more of a fraud was that half the time I didn't even think in terms of the word 'nature'. More often I'd be thinking of a specific place, some amazing, sigh-inducing landscape or a cool, twisty tree, or a small creature or squawky bird I spotted while on a walk in the countryside or in some meadow or park in my neighbourhood. And even if there were those who'd be empathetic, who would hear me? I often felt too conventional for the pagans, too esoteric for the hardcore wildlife tribe, not deep enough for the deep ecologists, not logical enough for the scientists, not 'listy' enough for the birder types, not enough of a 'green thumb' for the gardeners. All in all, I felt invisible, ignored by the cliques, and that I was becoming ill and needy with the desire to be heard by them. I struggled with the pain of being overlooked and of falling through the cracks. But I was also sick of it all, sick of the anxiety. This was no way to live, I realised, if I wanted to hang on to my sanity. It was time to just do the thing that I secretly longed to do: to actively seek to enter a world that co-exists with the visible one, a world of signs and portents; and to experience this land, my home Britain, as the indigenous people who I'd met in the far-flung places of my travels had experienced theirs, and to let the rest go.

The possibilities were too tempting to ignore. For what might happen if I embarked on such an adventure? What might unfold if I were to step outside of the box and wander and flirt with the land in a spirit of playful experimentation?

What was I seeking, anyway? A more intimate way of relating to the earth? For the land to guide me and see into me and speak to me? For magic to unfold before my eyes? For the gods to leave giant Post-It notes for me in the sky? Whatever I was reaching for, I craved communion. I hungered for it. I'd always loved to roam, but now I wanted to roam with a juicier intent.

Still, I was no land-whisperer, no expert natural navigator, no shaman. I wasn't rooted in a single cultural tradition. I was just a woman striking out on her own. How would I set forth? What was the plan?

This idea of throwing logic and order to the wind and letting my spirit and the land be my compass was all well and good, but I had to start somewhere. Where would I begin? Where would I go? This wasn't a 'from one side of the country to the other' kind of thing. Then again, maybe I would leave it to serendipity and the mysterious dictates of magic. For if I was going to do this, I'd need to enter fully into the spirit of my endeavour. Anything less would be a tepid charade, an exercise for my mind and not my heart. And I wanted what my heart wanted. I wanted to travel lightly too, with some levity.

At the outset, I held on to one thing: I had another intimate experience of Otherness – my own. I was British by birth, Indian by descent, Canadian by upbringing, South African via my parents' birthplace. I was always going to be an outsider, so this journey would be just one more facet of my outsider-ness. The wound of living in the margins was something I carried so deep inside me, was so much a part of me that I barely spoke about it. It was there though, everpresent. So in seeking the wild unseen, in a way I would be attempting to make contact with friends and allies. That's how I saw it, anyway. So what really did it matter what people thought? I had nothing to lose and everything to gain.

# Beginnings

My journey began in childhood. How could it be otherwise? We'd landed in the midst of a blizzard. A total whiteout: my welcome to Quebec, my new home. Have you ever experienced a blizzard? It's driving snow, an onslaught of white, stinging pellets from all directions, nature run amok. In Inuit mythology, blizzards were believed to have been created by Sila, an elemental spirit, as a form of punishment. If this is so, then Sila didn't show us newcomers much mercy.

In any case, it would have been hard to imagine a more startling contrast to the short life I'd led before in a south-west London suburb (where I was born on the winter solstice eve, as it happens). Nature, the outdoors – whatever you want to call it – hadn't been a big part of my life in my first six years. In contrast, my mother, who like my father is Indian and from South Africa, moved to the bush as a child for a time. This meant run-ins with poisonous snakes, and shimmying up the trees in which said snakes liked to sun themselves to pluck fruit. The wild was also in the snarling dogs she avoided on dusty back streets, and in the fishing and the foraging she did with her brothers and sisters (albeit out of need). My father's family had been more 'townie' but he had lived on a farm as a child back in South Africa for a bit too. His own father was a policeman, although his powers only extended to the Indian community: under the racist regime, unless you were white you couldn't actually police the upholders of apartheid.

What I'm trying to say is that nature was very much a part of their young lives – just not in a 'tramping around Dartmoor *en famille*' kind of way. But as newcomers to England they had more pressing concerns, like adapting to a country only marginally friendlier than the one they'd left behind. I can't pretend it doesn't make my heart ache to think about what they must have put up with in their early years in a chilly London. In those days the signs in houses for rent still said things like 'no blacks or dogs'.

They came to the UK because my father, who was active in the anti-apartheid movement, didn't want his kids growing up in such a toxic environment. That he was able to leave at all was down to fate. A kindly tutor of his developed a brain tumour and had to come to London for surgery. Happily the tutor recovered, decided to stay on and sponsored my father to come and do his MA, which then became a doctorate. At least, this is what I've gleaned in conversations with my mother, since my father is no longer alive. At this point in the story, my mother – still wedded to her saris and still sporting a red dot on her forehead (whereas now she pretty much lives in senior-friendly, catalogue-ordered linens and joggers and hoodies and is bindi-free) – had three little girls and in her scarce spare time sewed wedding gowns for eager brides.

My overriding memory of those first few years in London was one of a vague sense of oppression, of a premonition of burdens yet to be borne because I was different from the other kids in my school (there were not too many Asian faces in Wimbledon Park Primary). Had I stayed in England, would I have ended up a little diminished, a little crushed? Would that have been my fate? Maybe. I'm just grateful that I got out when I did. No one had been mean to me – yet. Or if they had, I've blocked it from my mind. And so it was with my innocence intact that, aged seven, I flew across the Atlantic.

⋆ ⋆ ⋆

I almost felt sorry for our host, cheery and stoic despite the burden of both blizzard and a collectively stunned family of four – myself, my parents and my middle sister. My big sister, away at university in deepest Devon, had stubbornly refused to budge. She'd only just flown the nest, so who could blame her? With the help of snow tyres and windscreen wipers, the car inched along and we made our way to our new home on the edge of a tiny village in Quebec's Laurentian mountains: Shawbridge. On the map it was a dot, two hours from the airport. I'm amazed it was on the map at all. It seemed more like a frontier settlement, dwellings separated by snow-drifts and snow-packed tracks.

Into this weird, wild winter wonderland I – the urban child – was delivered, agog. Our first home was a long, alpine-style chalet. Behind it was back country: acres of fir trees, branches heavy with snow, and more snow-drifts. I now lived in a snow globe! I spent the first few days and weeks gingerly exploring, skidding, plunging thigh-deep into the snow, awkward in new snow-boots, marvelling at the zips on my snow-pants – waterproof overtrousers. 'Pants', I soon came to learn, were never ever knickers but some kind of trousers or even jeans.

Behind our house, I discovered snow dens, deep squares carved out of the snow, each filled with a booty of snowballs. This was how local boys, invisible in their fur-lined hoods and with eyes peeping through balaclavas, played. I wasn't sure if they were friend or foe, and they seemed part of an alien tribe to me, so I'd watch them from a window and was careful to only show myself when they'd had enough and left. Once I peered into one of the dens and fell in. Terrified, I'd scrambled my way out in a blanket of silence. The snow muffled everything.

I had a toboggan and was soon hurling myself down any slope I could find. Ski-doo rides were extended as a neighbourly welcome. Clinging on for dear life to the dutiful teenage son of my dad's colleague, I sped off along a white carpet, silver spray flying as I dodged low-hanging branches.

It was as if the slate had been wiped clean, and I'd been set to 'reboot'. That feeling of oppression, the barely conscious fear that the world might not allow me to flourish: it was as if it had never been. I was still shy, lacking in confidence, not very sporty and ultra-sensitive – sadly, geography couldn't eradicate the less useful parts of my personality – but the landscape was a game-changer. The space, the nature and the quiet were exactly what an inquisitive, imaginative, seven-year-old needed. There was much to savour and I could do so at my own pace, without feeling jostled. The magical, the wondrous and a hint of something more and very big that the silence held – they now appeared to live on my doorstep.

I'd often stare at the ski slopes in the distance, watching the human ants descend. Could this really be the view from my loo window? At breaktime in school, I helped pat down igloos or hurled snowballs, though I was careful not to actually hit anyone. Some days I gamely flung myself to the ground and carved angels in the snow alongside my classmates. It may have been shockingly cold – minus 40 with the wind-chill factor – but the sunny, crisp days (of which there were many) dazzled me. Still, in that cold you don't stand around doing nothing. You keep moving if you want to survive.

With all childhood memories, when you look back, it feels like a dream. The difference is, it felt like a dream back then too. In that little village, in the countryside, I never once felt like I stuck out. There was a sweet guilelessness to the place and the people and it gave me a sense of belonging.

I was learning a whole new culture, the culture of francophone Quebec and the culture of cold – that's what nature was, this astonishing winter world. I didn't think much about wildlife. There were no moose or wolves in our bit of the world, and if there were any black bears they'd have been hibernating in the wilder spaces of Mont Tremblant National Park, a winter hour's drive away. White-tailed deer and red foxes lived in the province – I only know this *now* – but I don't recall seeing any.

Slowly we adjusted. My donnish father got stuck into his new job as a psychologist at a home for wayward boys in the foothills of the mountains and donned snow-shoes when the weather called for it. What the locals made of a bearded Indian in snow-shoes, I'll never know, but one day he strapped them on and headed off to rescue my sister, stranded on the way home in yet another blizzard. I can still picture them battling through the wind and snow, my sister's long curly black hair like fronds of frozen seaweed. Canadian winters are harsh, but there was no disputing the old-fashioned kindliness we experienced in our small, scattered community. My culinary vocabulary expanded to include tourtière, a kind of meat pie, and 'sugaring off' (a tradition of eating freshly tapped maple syrup on snow cones in the forest) along with blueberry pie, toasted Cheez-Whiz sandwiches, root beer, Del Monte puddings and the coveted Pop-Tarts.

After six months we moved to Saint-Sauveur, a neighbouring village. Now it's a popular ski resort, but back then it was quiet and some of the roads were still unpaved and sticky with mud and stones. I made a new friend, a tiny, dark-haired loner named Michelle who said little but who introduced me to the joys of berry-picking in the fields. Once we'd eaten our fill of the wild fruit and had filled our Tupperware boxes, she'd lead me up into the muddy slopes of the mountains, which come winter would be transformed into sleek pistes.

I couldn't name the summer scents that carried on the breeze or the butterflies that darted round us or the wildflowers we rudely ignored, but I do remember the sweet peace. It amazes me now that we never ran into anyone else, and that at aged seven, we were allowed to roam free like that. We felt perfectly safe – those old mountains may have dwarfed us but there was something affectionate and watchful in their presence.

At other times we storm-watched from the safety of my friend's home. Great forks of lightning would bear down between the mountains and we'd sit in the front room with its wall-to-wall windows, awed and terrified. Only when the light show ended was I allowed to head home. My mother would often be furious. Why had I been out for so long? Why were my boots so muddy? She had forgotten her own childhood in the wild, surrounded by her brothers and sisters, and now she longed for city life.

That meant we soon moved to Montreal, or rather a suburb of the island city. Winters here meant dusting off cross-country skis for morning treks across the local park under the maples, shorn of their leaves but bursting with sap that'd morph into the sweet syrup. Or it meant pulling on skates and dancing with the cold as best we could.

We had a house with a big garden and, even better, the St Lawrence River flowed past the end of the street. The bike path along the river led to what is now a beautiful bird sanctuary, Heron Island. Here the Lachine Rapids churned away, the waves leaping against the rocks every second of every day until winter set in. In the spring thaw and summer, it was an instinctive thing to cycle to the rapids, fling my bike down on the rocks, find a smooth one to perch on and daydream for hours on end. Often, I thought about the future. What would become of me? I wanted an adventurous life very much, even though I knew of no adventurers who looked like me. I was grateful for the rapids and could lose myself in their hypnotic rush and power. I felt lucky to live so close to them. Sometimes I'd shut my eyes in the sunshine and just feel nothing but *good*.

Other times I'd be worried and distracted. Things could be fraught at home. My father was no longer living under apartheid, but the fire of injustice still burned in him. Academic and professional though he was, the slights still hurt. When we had people round for dinner, he came out with angry, drink-fuelled, political tirades that left all of us on edge, dying with embarrassment. Other nights, he'd put on his Ravi Shankar records behind the art deco doors of our living room and sip whisky, the picture of loneliness. He sought solace in his garden and in his books: Alexandre Dumas, Robert Tressell, Krishnamurti, the Upanishads, Wilbur Smith – nothing was off limits. In his spare time he fished down by the rapids or on a patch of land he bought out in countryside in the Eastern Townships.

My mother was stubborn, loyal and she quickly abandoned her saris for Canadian 'pants'. A typical Indian mum, she cooked, she fed, she gardened alongside my dad, she listened, she championed, she kept her mouth shut, she cocooned us in maternal warmth. She surfed my father's temper stoically unlike me who, while I loved them both, liked him less. I'd feel my stomach lurch when he got angry or when I'd hear him unscrew the whisky bottle.

It was as a way of dealing with tension at home, or because I had things to figure out, that I'd head to the rapids, treating them as a confidant. If I had been asked at the time – and no one would have asked because no one talked about this stuff – I'd probably have said that nature wasn't for me because as far as I could tell lovers of the outdoors were a particular breed. They seemed happy to rough it (at that time I was even squeamish about walking barefoot on the grass), could read a map and compass, and were keen on camping, good at swimming in lakes, and pale-skinned. I was none of these things. But as a grown-up I got drawn in, through my travels and the experiences I'd had in the wild places I visited. All of it made me eager to go deeper still, to explore what it might be like to feel a spiritual bond with the land. I reckon the wanting had always been there, in my DNA, lying dormant within me like a seed waiting to burst forth.

# To the Oracle on the Sea

You have to start somewhere, and I start by stumbling on it during a random online search. It happens a few weeks after I spot the image of a labyrinth at Warren Street tube station, looking for all the world like a cryptic clue in a Dan Brown mystery. When I discover the World-Wide Labyrinth Locator, I feel a whoosh, a rush. At a glance it's as if a tribe of giant misshapen snails have gone walkabout.

Those peering at me from my screen are made from all kinds of things: there's one in the Karoo desert, made of cacti; another that circles quietly round a rondavel, like a giant anaconda. 'The Chalice of Love' in Australia is foresty green, made from thyme and presumably edible. There's a strange three-dimensional fantasy in Catalonia fashioned from wood and planks, like a Gaudi of the forest.

I get pulled down wormholes that bring me closer to home and I pause when I come to one on a hillside slope. It's in Cornwall above sea cliffs, on a farm owned by one lucky and quirky human. 'Walk it and lose yourself in the mysteries of time,' I read.

I just know, the minute I see it and before I've even formed the thought, that I'll go there. First, because it's by the sea. The sea always feels like a homecoming, replenishing, like a date with a briny divinity; I long to be near it – in my head the sea is often a sinuous 'she' – even though I can't tread water and am afraid of riptides and killer surf. And then there is the labyrinth itself.

There is something about it that fires up my imagination. You might simply call an outdoor labyrinth land art, but for me it teeters on the arcane and the mystical. It brings an Otherness to the landscape – and I crave that. It's like a portal: I can enter into an altered relationship with the landscape in a single footstep. I'd had a taste of that kind of strangeness as a child. In the big field of tall grass behind our home in Saint-Sauveur was a maze. How it got there is anyone's guess – for all I knew, little green men could have laid it down. It wasn't the same as a labyrinth, for there were dead ends in the twists and turns, but I'd often walked it dreamily after schooldays in springtime when the snow had melted.

I decide to contact Caroline, the owner of the Cornwall labyrinth. She has a longhouse on her 70 acres of land, part of which is a coastal nature reserve. And half a mile of her own beach.

'Just whizzing off to set up the Festival of the Sea in Looe!' she says, sounding flushed, when I get hold of her. 'But it's all yours for three nights.'

Great, I think, rubbing my hands together and trying gleefully to imagine what it will be like to have that great space to myself.

I sit down and do a bit of labyrinth research, and discover they have appeared on ancient rock carvings, that monks have walked them for contemplation and serenity, that they've featured in Greek and Hopi Indian mythology, and that some interpret the Aboriginal Songlines as labyrinths. People have spoken of life-changing transformation that came about from walking a labyrinth, and some walk them as mini-pilgrimages. I read that they're found on walls in hidden caves, on beaches, at the foot of mountains, in spirals of shells, on cathedral floors, in galaxies, hurricanes and those miracle-like fossils called ammonites, even in the hemisphere of our brain. And London's Victoria and Albert Museum has in its hallowed hall of blingy jewellery a labyrinth of precious gems (if you don't believe me, it's on the first floor).

Everything I read about labyrinths – often maddeningly vague – leads me to believe that you are meant to ask a question on the threshold. A burning question for which you require an answer. That sounds good to me. I have all kinds of burning questions. But who would I be offering my question to? I quite like the idea of a labyrinth possessing oracular powers, though. I'll go with that.

⋆ ⋆ ⋆

It's a sweltering hot day when I take the train down towards Liskeard. Across the table from me are a grey-haired couple, staring at me as if I'm the specimen on a petri dish under a microscope. I know that look, the one worn by people who are unused to being up close to someone of a different colour or even culture. Laughable to a Londoner. It's a discomfiting thing being jolted out of your cosmopolitan ease, in moments like this. A copy of the *Sun* lies between us and I turn away to stare out the window and contemplate the solstice. In London, among my circle, midsummer is practically a competition, people outdoing themselves to be the most devout, most honourable of Gaia worshippers, to attend the most wildly expressive ritualised gathering, to let everyone know that their way of attuning to nature at her peak, when the light is at her zenith, is the most profound. 'Will you be at Stonehenge, Glastonbury, Avebury? Will you be chanting, singing, making prayers, lighting a fire?' To not celebrate feels like a marker of extreme, un-green uncool. I am as bad as the rest, off to walk a labyrinth on the solstice morning. But I cannot escape the feeling that the day will turn the act into even more of a ritual.

On the train I sigh with pleasure. I adore – even feel a strange ecstasy – watching the countryside rush by. It's always been this way. Maybe I'm easily pleased, but there is something in that promise of a green arcadia, those haystacks, the frilly hedges, hills embroidered with broccoli trees, flocks of sheep… all of it makes my heart sing. It is my every fantasy of the British countryside come to life. I once tried to explain it to my friend Francis, who is a keen walker: 'It's just so exquisitely, quintessentially British! I know I live here, and I was born here and am a Brit, but growing up somewhere else I can also see it all and appreciate it the way a tourist must.'

'Well, that's good,' he'd said. 'Because most people who live here don't give a toss.'

I knew that last bit wasn't true but I was pleased that my love affair with Britain's countryside had begun to flourish. It wasn't always this way: I used to turn my nose up at life beyond London. Partly because back then, I didn't know anyone who lived outside of the city. Abroad was always better, more exotic and exciting, wilder and often more familiar too. Until the scales began to tip a bit more in my birthplace's favour.

At Looe station, there is no sign of life, and the heat is such that it feels like the air has been sucked away. It's all very Deep South and a bit disorientating. Yet it's hard to feel anxious in the heat. I find the Looe Valley branch line tucked around the corner, looking clapboard sweet and a bit Deep South too. There aren't many people around. Where are all the tourists? Eventually the train chugs in and I ask the conductor to please stop at Sandplace. I am so terrified that he'll forget and I'll miss my stop that I sit near the door, ready to leap off if need be. I can hear the train driver talking to him. 'Sandplace?' he snorts dismissively. What is wrong with Sandplace, I wonder?

We get going and cut through valleys snuggled between a dense tree canopy. The sunlight streams in nicely and we ease past tiny platforms that look like they belong to another era. One of the stations is Causeland, where ghostly sightings have been recorded. Then, at last – though in fact it hasn't taken long at all – we pull into what must surely rank as one of the tiniest train stations on the planet.

Standing by the gate, on what I'm guessing are the fringes of this tiny hamlet, is a tall, white, dreadlocked man in shorts. When I get closer I see that his feet are bare, gnarled and twisted like exotic tree roots. The last time I'd seen feet like this was when Mother Teresa, who I'd met once in Calcutta, laid a hand on my head and blessed me. (It was her icon I'd carried up with me to the Pyrenees and clutched in terror that strange night on the mountain.) I'd looked down, I remember, when meeting the tiny woman and had been slightly shocked at the sight of her crooked toes.

Anyway, the man's dreadlocks are piled above his head in the form of an elaborate sculpture. His name is Martin.

I know this because Caroline had prepped me on the phone. 'He's a caretaker on the property and he's as garrulous as a clam,' is what she'd said. I have been forewarned. Our exchange goes something like this:

Me: 'Hi, I'm Jini. You must be Martin?'

Martin, clearing his throat and looking at his feet: 'Yes.'

People I meet for the first time in the countryside often look at me a second longer than they need to. It's a different look to the train look, more wholesome somehow. Sometimes it makes me feel like a painting in a gallery – not always a bad thing, but it depends on your taste in paintings. I understand the curiosity, but I don't want to be reduced to the exotic specimen from Planet London. I don't have to worry about that with Martin. I don't have to worry about making polite conversation at all. Instead, I lap up the breeze with the window rolled down, and around 20 minutes of country lanes later we turn off onto the farm. Suddenly, as far as the eye can see, are the sea and the horizon. I can almost get a sense of the curvature of the Earth, it stretches out for ever. What a place for a labyrinth, I think.

The farm is midway between Seaton and Looe. Caroline built the circular, winding path after her partner had died. 'I was going through a difficult time and went to see a clairvoyant. She told me to go to the Rocky Valley in Cornwall. So I went and came upon a symbol of a labyrinth on a rockface. It felt like a sign,' she'll tell me at the end of my stay as we sip Cornish blush cider and stare out of her big windows at the sea. 'The labyrinth has brought something special to the place, a protective energy. I don't understand it, but I feel it.'

Feelings are like that: ungovernable, inexplicable, at odds with the rational. I like that Caroline is a reader of signs, because I like to think I am one too. Besides, I'd done that walk myself once – it was on the other side of the county, in Tintagel through ancient woodland along the edge of a river – and it was eerie and full of shadows. I don't recall if I'd seen a labyrinth symbol in the shale or if it now only looms in my imagination.

Here there is no eeriness. No electrosmog either, and without it the peace is something I can touch. I don't have to meditate; the land does the work for me. It's not an absence of noise but a presence that soothes. Woven into it is the birdsong, the breeze, the whoosh of the sea, the distant hum of boat engines, everything *local* and in its element. I can exhale at last.

'If you need anything, there's no phone signal but we're not far away,' says Martin from the door of the longhouse, pointing vaguely to a winding path before disappearing up it. On the way here, we'd stopped at a farm shop and so I have plenty of provisions. An empty house with a full fridge, and the sea – I want for nothing. I pace around like an eager puppy, poking my nose into the rooms, practically swimming in the space. For a moment, a familiar pang of loneliness creeps in, but I remind myself that the wrong company could jeopardise the whole endeavour – better to be alone. Anyway, I have the sea for company. And a labyrinth. On the kitchen table, I spot a booklet: *Small Pilgrim Places Network*. It is full of 'liminal places on the edge of mystery'. I like the wording; it validates my impulse to come. The labyrinth is listed as the only Small Pilgrim Place that is 'open to the four winds'. I like that too.

I think of all the things people have called me in the past – 'woo woo', 'hippie', 'eccentric' – and mentally I thumb my nose at all of them. Life is too short to care what others think of you. If I live in the cracks, if I fall between camps, if I have a foot in a few worlds but don't fully belong to a single one and am part of no clique, I'll make the outlier world my home, a desirable place to be, I tell myself. Others can find me here if they want to.

With new resolve, I decide I'll visit the labyrinth when I choose to and not be ruled by the solstice. That means I want to see it right now, this very minute, so I open the sliding doors wide and step into the sunshine. According to a map Caroline has drawn, the path to it is beyond a garden gate that takes me through a field of waist-high grass. I try not to think of adders as I walk it. The path is lined with foxgloves, like a guard of honour. It leads into a meadow full of wildflowers, and if I squint the effect is kaleidoscopic. The labyrinth, on a slope looking out to sea, is easy to spot. It looks like a landing pad for a spaceship. I approach it tentatively, like one might a wild creature or a UFO. Because I've seen it in pictures and now here it is looming in front of me, it takes on an unreal quality. The grass furrows are distinct and well maintained. It is, I am to learn later, 60 feet wide.

The wonder of the labyrinth is in its relationship to the sea. Straight ahead there is nothing but the waves; to the west, the curve in the landscape that is Looe. Just off the mainland is Looe Island, which was once owned by two sisters. Adventurers and bandits and religious recluses hid out there in times past. It might be owned by a wildlife charity now but it feels mysterious, like an Enid Blyton island, begging for a nocturnal visit.

Attached to a fence that borders a disused part of the South West Coast Path are some panels: on them are labyrinth lore and esoterica. The labyrinth is a symbol of great antiquity, I read. Coins from third-century BC Crete had the designs 'struck' on them. Hundreds of stone labyrinths were once built in Scandinavia in coastal areas, thought to be created by fisherman eager to trap trolls. I learn of the ones on the Isles of Scilly and am gutted not to have known about them on my one visit there. But it is the more mystical references that catch my eye: that there are always seven circuits in a classical labyrinth; that seven is a magical number linked to the colours in the rainbow; that the human body has seven energy centres or chakras, a Sanksrit word first referenced in the Vedas, the most ancient of Hindu texts. I am secretly delighted to find all of this in the midst of a nature reserve. Caroline, who has posted the information, clearly isn't worried about being met with cynicism and criticism. I admire her fearlessness. I can learn from her, I think.

The last panel talks of the oracle-like powers of the labyrinth, its ability to act as a lightning rod for the answer to your question. And it has to be a heartfelt one. There is a ritualised way of asking too: you have to ask at every left and right turn of the labyrinth. Complicated. Maybe you get into a meditative state of mind while walking and that helps you to find your answer. Maybe *you're* the oracle. Maybe as you walk, you peel back the layers and grow closer to your own nature, while out in nature. I don't really know.

But what is my question to be? I reflect for a moment: I can think of about five, and four of those I'd be too embarrassed to admit to. But there is one that keeps coming up. I have this yearning to connect with the Other in the land. I'm hungry for more, for magic. A childlike intensity of experience. I am not sure if I am seeking a communion with the Divine, with God, with a landscape deity, with the spirit of Mother Earth herself or with none or all of the above. *What am I seeking?* I ask. I take a deep breath and a big step forward with my clunky walking boot.

Some 20 serpentine minutes – and many perspectives of the sea and the meadows – later, I reach the centre. In it is a standing stone. It is not some prehistoric land mystery, but a gift to Caroline from friends. No answer to my question, though. I walk back out and I hear nothing. Have I not lingered long enough? Have I been too distracted by the sea views? I walk back in, and sit in the middle on the uneven ground with my back against the stone and stare at the sea, now a stony grey-blue and it tells me nothing.

I think back to when I'd told my friend Francis I was coming here and he didn't get it. 'But why?' he'd said. 'Why do you want to walk it? What's the point?'

'Because…' I'd had to think about it. Why *did* I want to walk it? What *is* the point? 'Because… I reckon the view will be quite something,' I'd said pathetically.

So is the view something? So far it's just a regular view. No losing myself in the mysteries of time.

I'd walked a labyrinth in Alfriston in Sussex with my friend Olivia once. It was behind a churchyard, cut into the grass between it and the River Cuckmere. It had been grown as part of a festival. Its motto was: *the beauty is in the walking*. I'd walked slowly and reverently but all the while I'd felt supremely self-conscious.

Another time I'd been to the island of Cres in Croatia. I was in the company of an eccentric biologist whose name was Dr Goran Sušić – the locals called him the Eco-God. He was dark and brooding. He was also a labyrinth maker, one who believed in fairies and tree spirits. He regularly retreated to the forest to meditate. I had come to visit his rescue centre for griffon vultures, though it was sad to see the birds in their cages, rescued after falling into the sea from their clifftop nests. In the end, it was the mysterious white stone labyrinths that I fell for – he'd built them across the island. The Eco-God believed they help to amplify our connection with the wild. 'They bring our souls closer to nature,' he'd said, sounding like a poet. He built them by hand, in the clearings of forests that he said were inhabited by spirits known as masmalić. On Cres, I'd walked the circuit. I'd sat cross-legged in the middle of it and tried to meditate, but it felt like a performance there too, because I knew the Eco-God was watching. I left, feeling enchanted by his creations and the forest, but none the wiser.

I'm disappointed not to feel anything here in this labyrinth by the sea, but I refuse to believe that it is just a fancy spiral. It represents a kind of sacred geometry, there is magic encoded in its DNA. If I can't decode it that is my failing, not the labyrinth's. I snap out of my mood. I am by the sea, I have an entire nature reserve to myself. How many people have this chance? I'd be mad not to make the most of it.

So I do and that day and the next, I walk along a trail made by the goats and sheep, across sloping meadows, the sea always within my sights, azure in the sunlight. The wildflower meadows are something. The colours are so vivid – the violets and the purples and yellows and the oranges – and now and again I feel a heady rush, as if the colours have been injected directly into my veins. Later, Caroline reads me a list of some of the flower names: 'birds foot trefoil, violets, speedwell, red campion…' The big blue spiky ones are viper's bugloss that grow everywhere on the farm. The plants sound as if they have been plucked from the pages of a fantasy novel or are the ingredients in a spell or part of an incantation: herb robert, hedge bedstraw, hart's tongue fern, cock's-foot and creeping bent. 'I'm creating a haven for the bees and the insects,' Caroline says.

When the footpath meets a wider track beyond the meadow, out of the blackness of a paddock emerges a giant grey horse with large feathery feet. It grazes my shoulder with its chin as it walks past. I want to hold on to it, to make friends, but the horse has other business and disappears off into another field. Still, I'm charmed by the encounter. A black and white goat with curly, spiralling horns appears on a footpath and gives me a friendly bleat.

I'm intoxicated by all of this – it's hard not to be – but in the space between the steps I take and in the silence something is missing. I can't quite put my finger on it. How can I be fully alive to the beauty here, I wonder? Not just the look of a flower or the sea or a leaf, but the life force within each. Things are happening in a register lower or higher than I am able to detect. I am not good at this, my radar is not honed. I don't have the ears of the deep listener. I have no such gifts. So I end up veering between flatness and rapture. And I struggle to keep loneliness at bay – the kind of lonely that keeps the most beautiful things and the most serene settings at arms length, behind an invisible glass wall. Maybe that is what has drawn me to the labyrinth – a desire for wholeness, for completion. For what is magic if not a spectacular climax?

I keep hoping Caroline will invite me for a meal, for a little contrast, a little company. But so far there is no invitation in the offing. Meanwhile, I stomp through waist-high grasses to let the adders know I'm here. I've never laid eyes on one in real life and am hoping I never do. At one point, I find myself walking to the bottom of steep wooden stairs that cut through a dark woodland, and come to a near-vertical drop. Someone has fixed a sturdy rope here, so I cling to it and shuffle down the slope on my backside, grateful there is no one to witness my ungainly slither.

At the bottom, the beach is pongy and ripe with seaweed and deserted (unless you count the seabirds). I pull off my boots and socks, and wet my feet in the sea: an icy current flows through me. But now that I have half a mile of sand and shingle at my disposal, all I can think is that it is noon, that I am dying of thirst and, most of all, that somehow I have to figure out how to get back up that cliff. Bear Grylls I am not. Besides, I have no idea when the tide will rise – probably not for hours, but I don't want to be marooned here when it does. I don't linger; I peer up at the slope and the rope, puff out my cheeks and blow the air out, and begin to haul myself up like a drunken Spiderwoman. With a final heave I crawl over the top of the slope and I catch my breath. And then I start laughing to myself: all I've wanted is to walk a labyrinth, and now look.

★ ★ ★

The days pass this way in a blur of sea and walks and swinging on the tree swing in front of the sycamores in the garden. I cast only occasional glances at the labyrinth, treating it as a suitor who has spurned me, giving it a wide berth because it has let me down, because I no longer have the patience to walk it, because it will not yield its secrets and because I did not get an answer to my question. On the last evening, the mist rolls in and the temperature drops several degrees, as does my mood. I get the wood-burner going, wrap myself in blankets, curl up on the living-room sofa and stare out as the fog rolls in, thick, eerie and mute. It's a pretty stark contrast to the sunshine I've basked in over the last few days.

I sit and idly half-daydream, half-doze and it is a while before I can allow myself to dwell on the labyrinth with a more forgiving heart. It has brought me here, but it has turned out not to be the thing for me here, I decide. Yet I have felt its presence even when I've not been anywhere near it. I've enjoyed a gluttony of solitude. I've had the sea to myself. No human has witnessed my sloth-like comings and goings. There've been no locals to collide with. And there has been complete freedom too from 'put-you-in-a-box' eyes that might have cramped my style, curtailed my freedom to strike out in any direction in my hesitant way. It's not just the fine weather or the meadows that have done this; the labyrinth, I am convinced, has had a part to play – like a benevolent spirit host. It has pulled me into this reservoir of beauty and somehow animated it. Maybe that is the nature of its subtle magic. But for me, it is not quite enough. It's a start, though.

# The Lost Spring

A few weeks later, I perk up. I'm chatting to a friend and she tells me about a treasure map leading to a secret, lost spring. She talks about it the way some might talk about drinking tea, like it is an everyday occurrence. 'Let me tell you the story,' she says one summer's day in the communal garden behind her flat in Notting Hill, unfolding a box filled with some high-grade vegan chocolate delicacy. We're on a bench in the shade of a monster London plane tree, trying to avoid the glare of the sun. The garden is filled with beautiful trees – mimosa, chestnut, hawthorn, ash and holly, a yew. At one end, two golden, privileged children in sun hats and thin dresses run round in circles. How can Charlotte afford to live here, I wonder? This neighbourhood, close to Holland Park's lovely Japanese garden and woods and peacocks, is mostly peopled by the haves. My friend, who in truth is more of an acquaintance, is nothing if not mysterious about her means.

Her story is about the man who excavated the spring about eight years ago. His name is Mark Golding and he is an artist and a friend of Charlotte. She doesn't so much tell me his story as point me to the actual written tale – I understand why, after reading it. It is a long, labyrinthine, passionate stream of consciousness. It deals with serious illness and despair and reads like a descent into the dark night of one man's soul, a journey from despair to redemption. It tells of the hours he spent in the woods engaged in hard labour, working with frozen hands, in cold, wet, mud-covered clothes, wading through swampy sludge, pulling up fallen trees and 'hostile and sharp shrubs' to clear away twigs and leaves and mosses. He worked in all weathers and at all hours of the day and night in the hope of finding the spring and its source. He describes himself as a 'wild woodsman, working on a wing and a prayer'. When he found it, he called the spring a miracle and spoke of it as a healing place. It flew under the radar but had a kind of cult following among those who'd come into possession of the map or the story.

I'm intrigued and ask Charlotte if she'll show me the map. I let out a gasp of childish glee when I look at it. The map is beautiful, hand-drawn, decorated with green, curling vines, footpaths and a sparkly mandala, a Sanskrit symbol depicting the universe. In place of the usual place names are more childlike descriptions: 'Peaceful Meadows', 'The Merlin Tree', 'Tangled Woodland', 'Olde Sheep Grazing' written in olde-worlde script. It is, I think, a work of art. A child's fantasy of a map – there is a 'Here Be Dragons' in squiggly script – and a grown-up's one too, if said grown-up had even a shred of imagination. The treasure is real, though: it is indeed the lost spring. At the top of the map are the words: St Helen's Spring.

'St Helen is the patron saint of new discoveries and lost treasure,' says Charlotte. 'She is based on the pre-Christian goddess Elen, who was worshipped and revered throughout northern Europe 10,000 years ago, back when these lands were covered in a thick boreal forest and our hunter-gatherer ancestors on this isle would herd and walk with reindeer on their migration trails.' Elen was, Charlotte says, depicted with reindeer antlers on her head. She was the reindeer goddess. 'The reindeer were so important for the life of the people: they gave them food, milk, hides, they knew the way through the forests, and where the waters were, the streams, all of it. The reindeer in a sense were a source of life. The animal was revered so the goddess was depicted as that animal.

'Her name throughout all of northern Europe in every language means "shining light"... and she is often connected to helping people find their way.' If you were going on a physical journey or needed help to get back on track in life, you could turn to Elen.

The spring, she tells me, changed her life. Bathing in it somehow helped Charlotte to harness her passion for water, specifically natural springs: finding them, mapping them, doing ceremonies in them and for them, and physically unclogging them too. 'I was given the treasure map in the way that I'm giving it to you. I went to look for the spring on my birthday, and when I bathed in the water I felt a golden energy, a presence – Elen's shining light. From then on, my path became clearer.'

She shows me a photograph of the spring on her phone: this is no trickle in the ground but a cascade that looks as if it's straight out of a fairy tale. It's shrouded by trees and it pours into a pool. It sparkles with promise, the sun glinting on the watery ripples like shards of gold. So where is it? There is an 'X' on the map to mark the spot. And a postcode too, making this a very 21st-century treasure map. Even someone as directionally challenged as I am can find it, I reckon.

I'm surprised and initially a bit disappointed when I see the postcode. I'd hoped for a more fittingly wild, windswept setting. Can a mysterious, secret springs exist in the suburb of a seaside town? The springs is hidden in a wood close to Hastings. *Hastings*, on the south coast. Actually, I think its location is proof that a landscape less ordinary doesn't always have to mean manoeuvres in a rugged, remote place. It can be camouflaged, hiding in plain sight.

St Helens Wood is at the foot of the High Weald, a landscape of woods and hills and fields and ancient trackways that bleeds into Kent, Sussex and Surrey. I'm not able to persuade Mark to meet up with me and share his tale first-hand – a shame, as he's into sacred geometry, a kind of arcane art where symbolism and geometry and the patterns in nature meet, and I'd like to know more about that. But no matter, now that both map and story are in my hands, I feel compelled to seek out the spring. Especially as everyone – not just my friend Charlotte – who has found their way to it speaks of being altered in some way.

I bring my friend Mandy on the wet, windy morning that I set out. A few months earlier we'd travelled to Iceland together and she'd humoured me and climbed a rock fortress that the locals told me was home to the Queen of the Elves. When I declared 'I want to greet the Queen!' Mandy didn't bat an eyelash, so I reckon she will be the right sort of accomplice. I have another reason for wanting to her to join me. Mandy has recently finished a course of chemo. Deep down I am terrified that days like this with her are numbered and I am greedy for her company. I also think a day in the woods will be a breather for her, take her away from her busy, urban life, from the responsibility of mothering a small child. More than that, I believe that St Helens Spring will be healing.

I'm not a strong swimmer or a wild swimmer but those times when I've waded into a river or stood under a waterfall on trips away, I've felt euphoric. A swim – just as long as my feet can touch the bottom – always leaves me feeling lighter in spirit, like I'm reborn. It has always been like this. I can't imagine what my childhood would have been like without the St Lawrence River at the end of my street or without the rapids to decompress by. Water was as much a fixture in my life as school or family. Even now, at the end of my street there's a tiny lake and a wood. I *need* to be by water. It's no coincidence, I'm sure, that the times in my life when I've been barracked by streets and dry land are those when I've felt most in a shambles.

I was once taken to a sacred waterfall in Queensland. Here women (often Western women) who wanted to have a baby came to bathe, for a female fertility deity was said to dwell in its depths. When I visited, I didn't swim but a file bulging with letters from grateful new mums around the world was thrust into my hands. Another time I was given an impromptu blessing – a shock dunking – in an icy waterfall called the Blue Well, or *Pozo Azul*, by a reformed drug dealer in Colombia's Santa Marta. The waterfall was sacred, he said, and I was half-terrified, half-exhilarated. Down in the valley of India's Living Root Bridges – a kind of Taj Mahal of the natural world – my guide left me by a natural pool to bathe but had forgotten to mention the python that also liked to hang out here. Fortunately for me, the snake stayed hidden and I entered into a blissed-out state wallowing in the shallows, lazily eyeing up the hobbity bridges. And the water I drank in Iceland? I could swear it tasted of life itself: I know if I lived there and drank it every day, I'd never get sick.

There's a whole tribe who seek out the wells and natural springs that exist all over Britain. The waters are hidden in ruins or deep in woods, or they flow beneath city streets – one of the biggest urban secrets there is. Charlotte is one of the tribe. She performs her ceremonies by London's waters and she isn't afraid to get wet and muddy. The previous summer I helped her to unblock a spring up on Hampstead Heath when we waded in in our wellies and cleared the muck with our bare hands. I'd joined her again there another time at the Goodison spring – a source of the River Fleet, one of London's largest subterranean rivers – to create a 'medicine for the water'. The medicine involved song and sprigs of yew and garden flowers and drops of gold oil, and the blessings of air, fire, water from the heavens and earth. How I'd loved the ceremony! We'd fed the potion to the spring. 'Now it'll flow into London's waterways and work its magic,' Charlotte had said. On cue, lightning had forked overhead, the wind rising and whipping our faces, and then the thunder came, a low, approving rumble.

So there are people who believe that blessings can heal ailing waters and landscapes (or at least halt some of the damage created by humans, for no kind thought or deed is wasted), just as there are those who have faith that clean, free-flowing waters and the spirits that dwell within them are capable of curing ailments of body and spirit. It's not just in times past that people did these things or cast their wishes into waters, divined their fortunes in water, left ragtag offerings of candles and feathers on branches and ledges near water. People are at it today, quietly, in the shadows. And now I'm one of those people. Charlotte has woven a spell and made this spring something magical. She's made me want to go to the High Weald.

★ ★ ★

It's a miserable and wet day – not very magical, and some kinds of water are harder to love than others – when Mandy and I follow the map to the ancient wood, by train, by bus and on foot. It's hardly an epic, glamorous journey. 'We can cut through here,' she says, taking charge and walking on ahead of me through the main entrance of what is officially known as St Helen's Park. The word 'park' sounds boringly tame to my ears. No treasure worth its salt would ever be accessed via a main entrance to a park – where is the magic in that? – and so I drag my heels. This is a special map I have in my hands, and I feel the right and honourable thing to do would be to follow it to a T. That's the whole point of a treasure map. I glower at my friend's back but don't want to rock the boat and so I trail behind her meekly.

Before coming here I'd read up on St Helen's Wood. It's a 104-acre estate with 'rare lowland meadows and abundant freshwater ponds' that support a 'rich and diverse selection of wildlife,' according to the St Helens Park Preservation Society site. Walking along a broad avenue lined with towering trees – ash, oak and perhaps a beech – it has the feel of a country park. I read aloud from the passage I'd printed off to Mandy: 'In summer the tall grasses and wildflowers are buzzing with insect life. Common and sheep's sorrel, cuckooflower and grasses are excellent food sources for caterpillars and butterflies including the orange-tip and green-veined whites. Meadow plants, the seeds that they produce and the insects that thrive on them, also attract birds and small mammals such as voles and shrews…'

That description – the very factual, unsentimental nature of it – feels worlds away from a treasure map and a lost springs and my own dreamy relationship with nature. But it speaks of the creatures without whom there would be no woods, no music and no container for the quite specific enchantment I am seeking. I wonder if the Preservation Society knows about the lost spring? I suspect not, because if they did the locals walking their dogs would know about it too. But when we stop and ask them, they smile tight, polite stranger smiles that don't reach their eyes, and shake their heads.

'Let's just keep walking,' I say to Mandy.

We cut through the wood, find ourselves by a stables on a back track, stop to pat the friendly colour-blind horses who peep over the fence to greet us, and take stock. 'I'm sure it's near here,' I say, peering at the scrunched-up map for the umpteenth time before letting my gaze travel over a field and a pond in which a heron sits, unblinking. I love how herons stand still on one skeletal foot for hours, dreaming heron dreams of tasty fish.

We cross a stile into the field, walk around the pool and then slowly make our way across a tiny stream, leaping from rock to rock and treading over branches we hope won't ricochet back into our faces, before finding ourselves in a copse.

Mandy pulls out her phone – shock, horror – and turns it on as if it's no big deal. She has a signal. 'Let's have a look at the GPS,' she says. I'm doing a good impression of 'The Scream' on the inside. *We have a treasure map. We don't need GPS.* We decide to shelter here from the rain, which is coming down harder now, and sit on logs and eat the snacks we've brought: the South African biltong (a nod to her motherland) and the hot rooibos tea from my flask (a nod to my mother's motherland).

We're no closer to the spring, so with a sigh I lean over her shoulder and we attempt to compare the landmarks on her GPS with the ones on the map. 'There's a church ruin that's meant to be near the spring,' I say but we can't see it on the satellite map, which we are both finding difficult to read. It's hard to tell one shadow apart from another and the ruin can't be where the little boxy squares representing houses beyond the wood's border are. There's no blue on the satellite map either. Would a secret spring show up blue? 'Does it matter if we find the spring?' asks Mandy. 'It's just nice being out in the woods.'

It matters, it matters a lot, I want to tell her. I've partly come to find the spring to show *her*, to inveigle her in the magic. I want her to connect with this enchanting story of the mysterious healing spring hidden in the wood as much as I do, and I'm disappointed that she doesn't. Besides, I really do want to find it. I am clammy and cold – we're both soaked through – but I'm feeling stubborn. 'Just a little bit longer, OK?' Mandy, with the patience that only a mother can muster, smiles and nods good-naturedly and continues the search with me.

We soon find ourselves in a meadow full of wondrous, twisty oaks, their branches spread wide. Gorgeous in the sunshine, I imagine, though right now the trees are smoky, ghostly, dripping, gnarled silhouettes. Just beyond the meadow on a footpath, we meet another dog walker, only she's friendlier than the rest. I show her the map. 'Oh, yes, I've seen those,' she chuckles, pointing to the 'Here Be Dragons' squiggled on it, although she's never heard of the springs or the derelict church marked near to it on the map. 'But I can tell you that one of those oaks –' here she points vaguely towards the meadow – 'is dedicated to Grey Owl.'

'A grey owl, you say?' I wonder if she's a member of the local RSPB.

'No, a man named Grey Owl. From Hastings, he was and then he went to Canada.' she says.

A man from Hastings named Grey Owl? We both nod in an interested way but she can't tell us much more, so I file this titbit away to be pulled out and chewed over later. Right now we have a spring to find. We take a tea break in a sodden field, and then give it one last shot. We walk up and down the footpaths, following streams in the wood that wind this way and that but do not end in a hidden spring. We fail to find the 'Weeping Willow Tree' marked on the map, and no matter how hard we look we can't find the 'Pilgrim's

Way' that leads directly to the spring either. We do find the sheep, dripping in a field, though. They look as miserable as we feel.

I can't help thinking that if we had followed the map from the start point, as we were meant to, we might have found it. Fortune favours those who honour the maps they've been given. In the end, we give up, take a bus into St Leonards, sit and eat noodles in a freezing-cold cafe, and head home, cold and deflated – or at least I am.

Back in London, I waste no time in ringing up Charlotte and whining. 'I don't understand,' I said. 'We looked *everywhere*.'

'It's not that hard to find. Maybe you weren't meant to find it today,' she says coolly. She always sounds so cool, so detached while sharing the most esoteric bits of information so generously. She is a hard one to figure out.

But if I wasn't meant to find the springs today, *why* wasn't I? Should I try again? Do I really want to? Is it worth the effort? It's a mad, pointless whim. I think about the train journey to Hastings. It took me to the sea, but it didn't feel uplifting the way a journey to Cornwall does. Of course, the Cornish coast has the edge: it's like a hallucinatory dream, so many shades of delight, what with the cliffs and the wildflowers and that glittering sea that I find so comforting. Towards Hastings the sea feels flatter, duller. It's a lazy, fleeting impression and I know it, but still it's how I feel. Maybe that's not the sea's fault at all but the town's. Hastings seemed to me faded and unloved, with the relics of brighter days – a fairground carousel and its worn rocking horses, cotton candy sold up a dodgy side-street – on display, pubs with rootless people loitering outside them. I don't really want to go back.

I suddenly remember Grey Owl, and decide to look him up. The story is fascinating, a little gem that has fallen into my lap. Grey Owl, I read online, was born Archibald Belaney from Hastings towards the end of the 19th century. He was drawn to the First Nations peoples and their way of life, and as a young man he travelled to the wilds of Northern Ontario. There he befriended a family from the Ojibwa tribe, who taught him their language and the arts of trapping and canoeing. He married into the tribe, only to then abandon his wife and baby daughter – classy. From then on he invented a colourful past. He told everyone who would listen that he was the son of a Scotsman and an Apache. That's when he began to call himself Grey Owl. History, of course, doesn't record what the indigenous people he encountered really thought or if they believed he was who he said he was. Maybe they decided to tolerate him or were touched by his respect for them. His not *othering* them. I try to remember the lessons in my Canadian high-school history classes. We'd learned about the *coureurs des bois*, the 'runners of the woods', toughened, maverick French-Canadian men who could integrate well into indigenous life, learn the First Nations languages and ways and the skills needed to survive harsh Canadian winters. They traded furs, and didn't have to answer to anyone.

Relationships were not clear-cut between the two parties. In the history books, written by white men, alliances were highlighted, conflicts downplayed: we barely learned of the missionaries who tried to convert the tribes, the traders who introduced them to alcohol, the human trafficking, the racism, the cultural misunderstandings. By the time Grey Owl arrived in Canada, the fur trade that had lasted nearly three centuries had just about come to an end, though he became a skilled trapper. The fur trade meant just that: the trapping and killing of animals, especially beaver, to meet the demands of fashion in Europe. I'd never read about animal rights in my high-school history book.

Though his relationships and family life sounded dubious – Grey Owl seemed to marry and abandon women and his children routinely – his great love, I read, was Anahareo, a Mohawk woman he married, from the Iroquois nation. The story goes that he hunted down a mother beaver, and his wife begged him to set her free. He refused but returned the next day for the baby beavers, which the couple adopted. So began his conversion from trapper to conservationist. Later, he even became a nature writer. He campaigned to protect the Canadian beaver, and saw it as a symbol for the Canadian wilderness. He became famous in Britain for a time, and it was only after his death that his real identity came to light. It turns out there's an exhibit dedicated to him in the Hastings museum.

The story of Grey Owl is a good one, but I'd much prefer to have discovered the lost spring today.

★ ★ ★

I put St Helens to the back of my mind. It's the height of summer, and my neighbourhood has become a giant orchard: fruit is hanging off the trees *everywhere*. I can barely walk two steps without filching an apple or two from a bough hanging over a neighbour's yard, or snapping off whole branches of elderberries to make compotes with – I love the bubbling-up of the mixture in the pan and the witchy feeling of making them. And there are rosehips in the garden. I toy with the idea of making candied rosehips from a recipe passed on by a forager in Leeds. She was black, female, Muslim, born in Kenya – not the usual gatherer of roots and shoots you hear about in this country.

When I'm not obsessing over summer fruit, I am trying to work and write. I am constantly trying to find ways to earn, to support my trips to the countryside, to plug the hole of my debts. I have rural land-ownership fantasies, and I have begun to crave big swathes of land with farm animals. I'm a dreamer through and through. Sometimes I think nostalgically of the house I grew up in in Montreal, with its river at the end of the street, and the mountain looking out onto the city. Sometimes I miss life in Canada. I desperately need time out but I can't afford a proper holiday. Instead, I meet friends for walks along the river in Richmond. Or in the park, where we point to the trees and shout their names in a kind of guessing game. Another pal is mad about flowers but insists on taking a photo of every single one we pass. I don't know what's worse, needing to know the name of every beautiful flower you come across or needing to photograph it. I join another friend atop the Tate Gallery, where we talk about the joys of magic mushrooms – she's tried them, I haven't. 'They *really* connect you with nature,' she says.

It's a hot, hot summer, but sometimes the humidity hits a peak and then it thunders down. On such a day I show an artist friend the Buddhist temple up near Wimbledon Common. She stops to admire the buddleia and the Japanese maples and I thank my lucky stars that I can come here whenever I want to. In Kent, I walk through the small wood owned by my late sister's husband. That same afternoon, he and I eat scones and jam and sip strong English tea while watching a cricket match on the village green near his home. He's started dating. He's finding his way to happiness again and I am happy for him. The summer unfolds like a song, a slow, lilting, lullaby that hits the top of the charts. I enjoy savouring this simple magic. There is nothing otherworldly or even especially wild about it. But it is perfect and it is mine.

★ ★ ★

Eventually I decide to return to the spring, right after I get some bad news that rakes up old trauma and sends me into a panic. Over a decade earlier, my sister and father died suddenly, within two years of each. Now a GP plants the seeds of fear in my mind. 'This is how it starts,' he says ominously, because my heartbeat is so low it may as well be on vacation. Freaked out and in need of a distraction, I book that train ticket. This time I really *need* a healing spring.

On the train down, I stare out of the window and wonder if I ought to write a will. My mood is funereal. In Hastings town, the sun is out, which gives me some comfort. I warily eye the queues at the bus station: a group of French exchange students sitting on the pavement chatting excitedly, older folks and locals who look like they have nowhere special to go. It's all so commonplace, and once more I feel foolish with my treasure map tucked away in my pocket. I'm relieved when the bus that'll take me into the suburbs arrives. The route is familiar, along the crowded beachfront, through the town centre and up and up through residential streets.

St Helen's Wood is a 10-minute walk from my stop. I walk down a suburban street and then an avenue filled with bigger houses and flamboyant gardens. I feel silly in my walking boots and rucksack, looking for all the world as if I'm about to tackle Everest. At a fork in the road, I carefully study the fresh copy of the map that I printed that morning. I cross the road, and this time I walk past the main gate and enter the wood via a track that runs parallel to it further along. It's not a particularly pretty or a wild-feeling track as it is on the very edge of the wood. It is more of a scraggly back alley, where I imagine stray cats hang out and weird loners walk their dogs. One side is bordered by a fence but the other is open to the wood. Almost immediately, I spot the Weeping Willow Tree from the map. Aha!

An old man is walking his dog up the lane, and passes me with a sideways glance. For a brief moment, I wonder how I must appear to him. I always feel a little more vulnerable when I'm on my own. Not because of a fear of danger; that's never it for me. It's something else. I never think of myself as different from any other person out enjoying the woods or some rural beauty spot – I'm just me – but I also know people in the countryside or even suburban parks don't often come across a woman with brown skin walking on her own. It's not a big deal, it's nothing really, not something worth my drawing attention to. Only, a little voice tells me that's not true. For my self-consciousness is real.

I admire the women who are black or Asian who never allude to their race, women who are so sure of who they are, sure of their voice, of their place in their world, or of their clear-cut ancestry that the word 'racism' doesn't even figure in their vocabulary, in public at least. The women who feel entitled and who inhabit a colour-blind world and just get on. Maybe they'd think I'm being paranoid. I think this is because this is often my *own* reaction when I hear people sounding off, people who I believe are seeing the world through a prism of victimhood. I've never been that, never felt that way. I grew up and was educated in an era when you never talked openly about feeling different or standing out like a sore thumb. It just wasn't done back when I was growing up in Montreal. It was just part of life, the way things were. Not that that made it OK. I still felt the things that I felt. I still felt vaguely uncomfortable in my skin and I was always trying to adapt, be more acceptably 'white', less ethnic. In school once we had to do a family tree. I was horrified, and with my parents' help – who knows what they were really thinking – I anglicised all those long, complicated Indian names of relatives that sounded strange and foreign to my ears, because I'd been conditioned by the society I lived in to view them that way. I even felt like this about the name on my passport, 'Sarojini', which I never, ever used even though I was named after an Indian poet and feminist. Did I fool my teacher when I showed her my family tree? My classmates? No one called me out on it and I was grateful for such small mercies.

The world may have been a calmer, more tolerant place then, but it was full of contradictions. In the city and province where I grew up francophones and anglophones were politically divided, so language rather than race was the dominant issue. And on the whole, the vibe in Montreal was a cosmopolitan one: Canada, after all, is a land of immigrants. But I would never have dared to share my thoughts about how I felt 'not good enough' because of the colour of my skin. And yet in other ways I was privileged. I had a nice middle-class upbringing and I wanted for nothing. Nothing was ever simple or straightforward.

Suddenly, I wonder what it must feel like to be a black man who enjoys going for hikes in the woods, any wood. That feeling of being conspicuous, of being othered. It's all relative, I suppose. When I'm done thinking these thoughts, I turn back to Mark Golding's story. I decided to bring it along with me today for inspiration, and now I look at a passage I've highlighted: 'The woods are ancient and living upon a series of hills and valleys, with streams interspersed, running through as arteries of the hills, carrying water, minerals and nourishment. A source of life-giving energy.' Time to focus.

* * *

A few minutes ahead of me, just as it appears on the map is the 'Pilgrim's Way'. I could swear blind it wasn't there when I came with Mandy, and yet here it is, unfurling like the Yellow Brick Road. It makes no sense at all and it's as if a kind of sorcery is at play. This track runs alongside a darker, denser stretch of wood. According to the set of footprints on the map, somewhere along here I need to veer off the track.

I start walking along it and almost immediately I spot what looks like an opening, some steps fashioned from an old tree trunk, or maybe it is stone. My mind feels absurdly sharpened, cat-like, as I climb up the bank of the ditch. In front of me is a fallen tree laid on its side, its two bald branches split like a wishbone. To one side of the tree is a sort of gully and what looks like an old, dank, near dried-out swamp. I walk over the fallen tree and deeper into the woods, and then suddenly another track appears. Again, I have the uncanny sense of it materialising only when I get up close. Like it's a hidden track, and you have to earn the right to see it and follow it. It runs parallel to a steep bank and as I walk along it I have a feeling that the woods is watching me, almost egging me on. It's beautiful here and still. So still that the hairs on the back of my neck stand up, just as they did when I was alone in the mountains before I heard that strange voice.

I'm straining my eyes for signs, and then an impression forms of a glade framing some moss-covered bricks up ahead. I hear it before I see it, that familiar rush that sounds like a song. And then I do see it! I can just about see the water tumbling down over a ledge like a silky curtain – into what, I'm not sure, but I've found it. I have the mad, wild feeling that some wood-dwelling spirit or naiad is exerting a magnetic force, pulling me in closer, welcoming me. I feel warm and protected.

And then I'm kneeling by the spring, astonished. It's beautiful – all of it, the setting, the flowing water. I try to take in everything I see, to commit it to memory. The water gathers in a small basin, rising up through the ground – this is the spring source. Later, Charlotte tells me she reckons it comes from a deep aquifer, and that on the other side of the hill is the church I've been looking for, so perhaps at one time churchgoers drank from these waters.

Someone has placed white feathers deliberately above the basin. A kind of ragtag offering, I suppose. The water flows from the basin and sings itself over the ledge that I spotted from afar, into a bigger, perfect, circular pool, big enough to plunge into. I can see now that it is lined with moss or lichen-covered stones, not bricks as I'd first thought. A corner of the pool is illuminated by a shaft of sunlight lasering through the trees. From here, the water runs down a stone-lined channel before disappearing into the woods and out of sight. Behind the springs is an oak, and nearby a spindly yew and, I think, a beech.

It's an enchanting spot, far more intimate and inviting than I could have even imagined. A Rumi quote comes to mind: 'What you are seeking is also seeking you.' I hope this might be true. To think that this spring had been clogged and hidden for who knows how long, covered in swampy filth and a mass of fallen trees and branches. It would have remained so were it not for Mark, driven and unstoppable, who cleared it with his bare hands, a shovel, a trowel and a scrubbing brush.

It's irrelevant to me that it's on the grounds of a derelict eighth-century church or that it may be even older than that. None of this crosses my mind, for I am gazing longingly at the springs and kicking myself for not bringing a swimsuit. *Should I…?* I ask a springs guardian I can neither see nor hear to protect me from prying eyes of stray passers-by, though I've not seen anyone since I've climbed over the ditch, which feels like a lifetime ago. And then a kind of feverish urgency comes over me. I pull off my jeans, my top, my bra and pants. I *have* to bathe in this pool. After looking left and right one last time, I plunge in.

It's icy and alive and in that instant, I'm no longer a thinking, society-bound human, but a human animal. I gasp with the cold and my audacity, for I'm not the skinny-dipping-with-abandon sort. I can count on one hand the number of times I've shed my clothes outdoors. At this moment, I feel closer to nature – to myself – than at any time since that strange week alone in the mountains. This simple act feels profoundly liberating and defiant and joyous.

I beg the water to heal me – I need it so badly and in so many ways. I appeal to St Helen, to the dryads of the woods, to anyone or anything that might be listening. I emerge dripping from the water and shivering and begin to pull on my pants… but then I think *too soon!* I've not let go enough. So I strip off and plunge in again. Every cell in my body is vibrating. There is the thrill of nakedness and reckless daring, mingled with the fear of being found. When I finally climb out I dry myself with my waterproof jacket for it's all I have, and dress quickly. I pull my socks on over my damp feet and lace up my boots and sit in complete peace in the glade. I shut my eyes and listen to the satisfying gurgle of the water, and for once no thoughts are running wild in my head.

And to think I'd woken up this morning in such a bleak and despairing mood.

Eventually, I gather myself and walk back to the huge meadow. It is so inviting in the sunshine, the grass a sweet blur of purple and yellow wildflowers. The oaks are here, stately and twisty like a gathering of elders – I am not sure which is Grey Owl's – and I find a bench and sit with my face to the sun and smile when the dog walkers and the locals pass me by. If only they knew such watery gold existed in their midst. If only they believed it possible.

I feel like a superhero, invincible and elated. I've done it: I've cracked it, I've conjured the magic. It's a heady feeling.

# Walking through Woods and Pain

When the student is ready, the teacher will appear. So goes the saying, and maybe that's why it was a whole, shameful, five years after I moved to the 'burbs that I finally twigged – no pun intended – that there was a wood at the end of my street. In my defence, it's a long street.

I'd come to this leafy, south-west London suburb synonymous with tennis and strawberries and cream because I'd had to. Not long after my parents returned to England to retire, after quarter of a century in Canada, my father died. My mum was on her own and she needed me.

At the time I was teaching English in wintry Tbilisi in the aftermath of the civil war – a short-lived adventure that involved waving postcards of the London sights at teenagers who adored Michael Jackson. (Sample student–teacher dialogue: 'Miss, we love black!') At least the Georgian natives were friendly. I left the craziness there for a different kind of craziness back in London. But sad though the circumstances were, leaving Tbilisi was a relief. No more bullet holes in buildings to shudder at. No more sinister-looking Cold War characters lurking in the shadows. No more lugging kerosene up six flights of stairs to heat my Soviet-era apartment block and nearly setting the place on fire. Sadly no more of the intense friendships I'd begun to form either. I hadn't wanted to teach but I hadn't known what else to do. It was early into my decade of despair, or as I call them 'the misery years'. But nobody tells you this when you're suffering: there will be a time when you will look back and the pain will morph into a dark, polished, prized saga, wisdom on a plate that you'll offer to a loved, anguished one in the hope of easing their pain a little. At the time, though, it was hell.

The decade before had been good. More than good. After university in Montreal, I'd studied in the south of France and when I failed to land a job after my studies, I moved to London and worked for an iconic book publisher. At first I was nervous and anxious in the Big Smoke, afraid to enter a pub for fear of having racist remarks hurled at me. I had this idea that London was full of skinheads, but that was all long gone by the time I turned up. Anyway, it wasn't the remark I was worried about so much as the embarrassment of the insult being witnessed by others. The shame of it. But then I relaxed a bit, settled in, found a shoebox room in a neighbourhood where the vibe was sunny affluence. Every morning I sashayed across Holland Park to work, as full of myself in my own way as the strutting peacocks in the park. Four years later I left the job to 'find myself', which in my case meant nearly dying of altitude sickness on a Himalayan mountain pass and volunteering at Mother Teresa's in Calcutta till the money ran out.

On a tip-off from an expat, I flew to Hong Kong and found a job working for another, far less glamorous, publisher. I made friends, fell in love with a tall Dutchman, got dumped and came back to Britain. That was when things turned sour. Losing your sister and father in the space of two years, when you have also become unemployable and are heartbroken and have endured a stint in a then-apocalyptic ex-Soviet country, is a lot to suck up. Somewhere along the line, I decided I would write an article and send it to the papers. What would I write about? My time at Mother Teresa's, that's what. 'Christmas Day in the City of Joy!' read the headline in *The Times*, wonder of wonders, when it got published four days before Noel. I was, briefly, ecstatic – I was a *writer*. My childhood dream come true. But I had no game plan and went into a tailspin. I spent the next seven years in a fug of despair and hopelessness, fleeing from ever more hideous temp jobs. When I wasn't running, I was grieving.

In south-west London, trying to make sense of it all – not just my father's and sister's deaths, but what felt like a total failure to make anything stick – I'd gravitate like a homing pigeon to Holland Park. It brought me back to the me I once was: the me who had loved her London publishing job, the me who had shone, even the me pre-grief. I went there every week, religiously. I'd exhale once I got through the gates, race to the Kyoto garden, throw pennies in the wishing well and pray grimly to the carp, wander around the Henry Moore sculpture and up and down the leafy, wooded trails behind the cafe. In this place, my pain was bearable. Why was I so stuck? I'd ask myself over and over. It wasn't fair. I was in a nightmare I couldn't wake from. I figured my life as I knew it was over, with the best years behind me. I may as well face up to it.

Little by little, my attachment to Holland Park waned and I discovered Wimbledon Common, a mere bus ride from where I now lived. I'd stomp moodily through the woods to the Windmill cafe, sip tea and weep and wail and plead with the gods. Where were they, anyway? Walking out my shame at having absolutely nothing to tether myself to – no job, no partner, no home of my own, no father or sister – was becoming a habit.

On my walks to the Common, I'd pass the tall iron gates to Cannizaro Park and one day decided to take a look. The park was hidden behind a stately country house hotel and was more of a landscaped garden with woods and extraordinary trees – the kind of park you felt you couldn't just turn up at in your slobby, holey leggings. At the bottom was a winding path where in season the rhododendrons flourished and the towering redwoods, birches, maples and horse chestnut trees arched or twisted and generally invited you in in defiance of the human code of formality the surroundings seemed to dictate. Every step I took was a prayer. Sometimes I'd wrap my arms around the branch of a tree and beg the tree for help. I was desperate. I prayed and prayed. And eventually things did change. By some miracle, I began to get work as a journalist. I was giddy with joy and relief. Through work I roamed and had adventures I'd never dreamed possible. I turned into a travel bore – the worst kind. If someone asked me how I was, I'd simply tell them where I'd been or where I was going. My identity was tied up in names on a map.

And then I began to ask myself what it was that was missing. Because something was. On my travels, I'd yearn to go off piste, away from people and into wilder places, sometimes in countries unpopular in the West. I wanted to meet the local medicine woman or man, not just for the magic and the mystery of their calling, but because on some level, I sensed they'd accept me because I too did not fit in. Such encounters would almost never feature in my itinerary, carefully cobbled together by a conventional minder or well-intentioned PR. And if they did, editors would too often weed out any references to them in the features I wrote, even when I pleaded with them. 'It's sentimental fluff,' said one breezily, his cut-glass, public school accent brooking no argument. End of. But, oh, the thrill of meeting someone who'd talk about the elements or the sky or earth or tree or a peak or waterfall as if it were a special friend or part of a clan of mysterious, sentient beings. In my book these indigenous people weren't sentimental – they had *vision*. My heart would leap and inside I'd be screaming *yes, this*.

But what exactly was I seeking? What was this thing that ran deep that I hungered for? It had to do with the natural world and with landscape but also something more. It wasn't a forensic thing. It wasn't a cold, detached, indifferent, objective thing. I walked and reflected and clutched at wisps of half-formed thoughts. I delved into the world of spiritual ecology, tried plant medicines and made friends with shamans (long before they became shorthand for a brand of consumerist, urban hipness – insulting to the genuine healers I knew). I dabbled, dipped a toe into arcane practices and took a vague, non-scholarly interest in the writings of nature mystics.

Around this time, I finally discovered the woods at the end of my street and the lake beyond it. I'd walk up there a few times a week and learned the names of the waterfowl from the information board. I got to know the ornery coots, the Egyptian geese, the moorhens and the mallards, male and female and the Canada geese with their exuberant runway landings and take-offs. Air Goose, I called them.

After my time by the lake, I'd turn back up the woods and head home, calmer and with a stronger sense of edging towards something. I just wasn't sure what. Or maybe I was too afraid to voice it, to say: 'I want to go deeper, I *need* to go deeper. To tread a more mystical path.' But in the end I had to. My parents, born into the sharp end of apartheid, had struggled and taken heroic, courageous steps so that life could be better for their children, so that I could choose my beliefs and my path, so that I could walk in freedom. To not claim this freedom would be dishonouring them as much as myself.

I had to tell myself over and over too that to want to steer clear of the cruel or even impersonal face of nature didn't mean I was living in la-la land. There were plenty of others who already told those stories. There were plenty who spoke out about the climate emergency too. I supported the protesters, I joined them, but the longing that I felt was the longing that I felt. I could not deny myself.

I could appreciate the physical beauty of the land and its creatures but I also wanted to be led by nature's invisible hand, the wildness you could not see. I realised I was happiest with a foot in both worlds.

# A Woman of the Old Ways

That wildness you could not see – well, some *could* see it. A shaman could.

I was acquainted with a white shaman from Botswana who'd been diagnosed with the 'illness of calling', as he put it, when he was 11. He was the son of a well-known anthropologist and a healer mother, and as a child had spent time with the San people, who had recognised his gifts. As an adult, he was trained and initiated as a sangoma, or diviner, and an inyanga, or doctor of traditional medicine who worked with herbs. He was quiet and modest to a fault – unusual in the world of medicine men in my experience. I'd studied with him once in Devon for a brief, enchanting week. I wasn't some special, handpicked apprentice – we were a motley group – but I soaked it all up, and by the end I began to feel less hesitant about airing the fascinations I'd secretly harboured.

It was the sangoma's sincerity, his lack of a need to convince anyone that he was who he claimed he was, that did it. On the day he had 'thrown the bones' African style, an Oxford-educated ecologist and a geographer – two men of science – had made a point of joining us. They were as rapt and as respectful as the rest of us.

At the end of the week, we fledgling, unexceptional students were sent into the woods before dawn. We'd been invited to enter into a liminal space, to see everything before us as laden with meaning: the butterflies that danced in the meadow, the swirling mist, the hazy, buttery yellow of the morning sun, the dogs that came sniffing along the path with their owners, the gentle rustle of the leaf-laden branches above us, the pre-dawn shadows, the soft soil beneath our feet and the moods we experienced moment to moment – all of it. And all because the liminal space is the one in which a shaman dwells.

Can you imagine living in a deep and constant relationship with the forces of nature? As I understood it, a true shaman does just that and believes that all living forms – humans, animals, mountains, oceans, rivers, trees, plants, etc – share the same fundamental life force. If you believe that at our core we're all moving energy – and physicists have told us this is so – then by a shaman's reckoning it is possible to communicate with everything. How? By entering into an altered state of consciousness. A shaman is skilled in the art of sensing and moving energies in order to bring about some kind of transformation or healing. That dawn vigil in the woods was intended to offer us a glimpse of what it would be like to attune to nature in this very special way. Well, it worked for me.

Soon afterwards, I read an essay by the American philosopher and writer Thomas Berry. It was an anthology called *Spiritual Ecology*. Berry's essay, 'The World of Wonder', put into words some of what I was aspiring to in throwing myself into these kinds of experiences: 'There is a single issue before us: survival. Not merely physical survival, but survival in a world of fulfilment, survival in a living world, where the violets bloom in the springtime, where stars shine down in all their mystery, survival in a world of meaning.' I wasn't exactly thinking *survival* – that was too extreme – but the world he painted was the world I yearned to be a part of.

Of course, not all shamans were the real deal. I'd met a few in the UK who seemed to suffer from an excess of self-importance, who used vague language and talked in riddles about their special ceremonies. Mystical didn't have to mean mystifying. In the rolling hills of KwaZulu-Natal province's Midlands, I'd encountered a Zulu sangoma who had a vast garden filled with over a hundred healing plants. He'd helped a former president find inner peace and performed divinations for film stars in LA, he'd said, showing off. He too had thrown the bones for me (I would always have trouble with money was the verdict) but he'd seemed distracted and so I'd nodded and privately dismissed him. In contrast, over the past few years I'd developed a close friendship with a shaman in London. She worked with clients around the world, and was as empathetic and as gifted as they came. She paid due respect to her mentor in Peru, with whom she'd studied the ancient healing arts, and who she continued to visit and study with every year. But she rarely wanted to talk about the intricacies of her work. She needed a break. She was a mate. We talked about the things mates talk about. We drank wine, we met for dinner, we gossiped... and it did not feel right to ask for her time to further my own journey.

Maybe it was a reaction to hearing too much about wild men of the woods – with one or two exceptions, men conquering peaks and deserts, men ruminating on nature and wildlife, their words precise and crisp and delivered with the irritatingly unassailable confidence of privilege, but more and more I craved a connection with nature through the messy, complex, passionate lens of women. Or at least women who saw nature as I did.

And that's how I end up in Herefordshire in the home of Elen Sentier, a woman who describes herself as a British Native Shaman. I'd met her at London's Southbank Centre, at a festival called 'Belief and Beyond Belief' where she took part in a panel. She'd worn a jaunty black hat with a feather in it. And she was over fifty, owlish in her glasses. If women are underrepresented in all things nature, can you imagine how the field thins out when you hit the big five-o? Elen was, she explained, a follower of the Old Ways and she lived in the Welsh Marches. 'The back of beyond,' she called it.

I was intrigued because she'd taken on the name 'Elen' in honour of Elen of the Ways. The same Elen Charlotte had talked about, the one who was the predecessor of St Helen, the guardian of the secret spring I'd visited. The Elen who was an antlered goddess and female spirit, the soul of the forest and of all wild places, a protector of the land.

Elen Sentier had spoken of Palaeolithic hunter-gatherers who worshipped the reindeer goddess. You could find reindeer rock art dating back 14,000 years on the wall of a south Wales cave. Elen of the Ways had appeared in a story in the Mabinogion, a collection of medieval Welsh tales. And among the circle of Elen scholars was another woman, Caroline Wise – the numero uno Elen 'goddess expert', according to Charlotte – but I'd already connected with Elen Sentier and, besides, I wanted to see Herefordshire and this was the perfect excuse.

I wanted to know what it meant to be a follower of the Old Ways. Was it something you were born into? Had she been born into it? Could I learn anything from Elen Sentier? She doesn't say no when I ask her if I can spend a few days with her.

⋆ ⋆ ⋆

On first impression – second too, to be fair – the Herefordshire countryside is a picture of pastoral bliss (a cliché, but true nonetheless), bathed in golden light. It is landlocked, and bordered by Wales, Shropshire, Worcestershire and Gloucestershire. Without the obvious charms of sea and cliffs, it's as far from an attention-seeking diva as a county can be. I also imagine it to be full of people who are indifferent to London and Londoners, and who can blame them?

The plan is that I'll stay in Elen's home for the few days of my visit, but when I arrive in Hereford my heart sinks as I see her propped up by crutches behind the ticket barrier, her hands gnarled from rheumatoid arthritis. She appears to be in pain, her lips set in a grim line. 'When I was young I studied contemporary dance!' she says, grimacing and, I think, wanting me to know that she has not always suffered in this way. Elen doesn't look well enough to be hosting anyone, let alone a perfect stranger. Still, she'd not rung to cancel and I am here, so I resolve to make the best of it.

In the car, we drive along untroubled roads that run like a ribbon of silk between the orchards and meadows and farm fields. There are sheep in the hills, little daubs of white. I imagine that the people who live in the houses hidden away here must be happy amid such golden surroundings. In the sunshine, I envy them. Soon – too soon, because sometimes I'm happier travelling than arriving – we turn down a long drive on an industrial farm: not quite the back of beyond I've been expecting.

Elen's house is in a cul-de-sac at the end of the drive, past some desperate-looking cows squashed into a barn. I'm shocked. How could anyone live in constant proximity to so much misery? Maybe she has no choice. Maybe circumstances have dictated her choice. Maybe if you wrap yourself in a mantle of magic, you can tune out the misery. Maybe, because Elen is a lover of animals and wildlife, she feels empathy for the creatures. Maybe hers is their only source of kindness, the only good vibes trained in their direction. Who knows?

I'm relieved when she leads me through the hallway to the garden. Elen once designed three biodynamic gardens at Hampton Court Palace Garden Festival and her skill, even to my untrained eye, is plain to see. She's created a wildlife sanctuary: in front of me is a pond with fat lily pads. At one edge is a cascading water feature, the water travelling through the 'flowforms' like a mountain stream. I can hear a sharp, rolling *krrrk*. 'It's a moorhen and her newborn chicks,' says Elen, lighting up. 'They're hiding in the undergrowth.'

In fact, the garden is filled with birdsong, like a summer symphony. And there is colour everywhere. 'These are flowering rush, these are crocosmia… roses, lady's mantle, pink, white, and magenta-coloured geraniums…' Elen reels them off. There are gawky wildflowers springing up and reeds and bushes, and oh, the trees! Cedar and hawthorn, holly, willow, hazel, oak, ash, yew and even a crab-apple tree are planted here. 'Look at the leaves, look at the fruit, can you tell what kind of tree it is?' she asks like a patient schoolteacher when I struggle with the beech and the walnut. As delightful as the trees are, my eyes are drawn to a mysterious, narrow, high-backed chair hewn from wood on one side of the pond. It is for an invisible 'ancestor'. At the foot of the chair is a skull.

There is another side to the garden, one in which the sun never reaches. Elen calls this 'the dark glade'. You enter it through a giant hoop, a kind of threshold. 'It's called the dark glade because it's full of the unknown. You might even say it's a bit creepy,' she says. Creepy or… bewitching. I clock the strange tree roots, what looks like discoloured sheep's wool, which Elen says is dried pondweed, another skull (is it real? I don't like to ask), a giant amethyst crystal, some dead wood, some tiny green dragon sculptures at the foot of a lopped-off tree trunk, and another of those mysterious chairs. I don't know what to make of these objects – I think that maybe they represent the unseen, the mysterious, shadowy side of life, or the Underworld, or maybe Elen is just expressing her artistic side – but I like the arcane touch they lend to the garden. It feels like a votive place, a shrine to the Other. Things are beginning to look up.

We sit on garden chairs and sip tea under the shade of a giant umbrella – we're still in the throes of a proper heat-laden summer – and Elen fills me in on her childhood. 'I was born on Dartmoor into a family of "cunning folk",' she says. Her grandmother on her mother's side was a 'witch from the Isle of Man' who could talk to animals and plants. Her father knew Annie Besant and Helena Blavatsky, the 19th-century founders of Theosophy, a movement that focuses on mystical experience and direct contact with the Divine. Her mother died when she was a toddler, and Elen grew up in a village on the edge of Exmoor. Here she was surrounded by healers and herbalists and wise women and woodsmen. An uncle would take her to the woods at dusk to see the foxes. They'd find a tree, maybe an oak in a glade, and sit at the foot of it quietly.

'We wouldn't light a fire. We'd sit still for a very long time. We'd wait and wait and then the animals would come right up to my uncle. A fox would come closer, and touch and sniff him. I'd sit frozen and then the fox's nose would touch me! Can you imagine? An owl would sometimes sit on his arm. I was entranced: to a child, it was like being in a fairy story,' she says.

Her uncle would show her spiders' webs in the autumn. 'He'd say we have threads like those spiders' webs that could reach out to every plant, flower and creature, only they were invisible…' Later, he would take her into the woods to learn about trees. Not tree identification, but something more enchanting and intimate: 'My uncle would say: "Reach out your hand and send your thread to the tree. Try to feel it from your heart." Then he'd ask me to do it without my hand. "Can you feel the tree sending a thread back to you?" he'd say. That's how I got to know a willow. I'd feel a thread coming back. It felt like a ping in my heart all the way from the tree.'

Ellen was never the only one who played these games. All the kids she hung out with did the same. 'We could connect with animals, plants, trees, water, fishes.' She calls it a 'kenning', a kind of knowing.

Can she teach me, I wonder? I am eager to learn these subtle arts. I'm more than eager: I'm primed, ready and *longing*. I hold my breath.

Elen shakes her head: 'It will take you a lifetime to learn, and even then…' She is not keen.

I don't know what to say to that. Does she not want to impart her knowledge? I know that I'm asking for a lot. There are rites of passage, serious initiations to go through before knowledge can be imparted. You can't just rock up and expect instant gratification. We aren't talking about party tricks. On the other hand, I have come all the way to Herefordshire for this. 'Can you at least give me some pointers?' I plead.

Elen trains her owlish eyes on me. There's a pause. I think maybe she's wondering how to get rid of me. And then, blinking: 'Listen. Be still. Get your head out of the way. Allow yourself to daydream.' I swallow the crumbs. 'Following the Old Ways is about learning to contact the land spirits. You can only do this once you've learned to listen. Then you ask them to show you what you need to know. Our Palaeolithic ancestors would have been very good at this – when they needed wild meat the whole tribe asked the spirit of the land, and the spirit of the animal they wanted to eat, for assistance. It was a sacred exchange.'

What she is telling me doesn't sound strange: the indigenous people I'd met on my travels spoke of their relationship with the animals in their midst in the same way, whether they hunted them or not. I think, too, of the adventurers I know, the ones who take film cameras into remote forests and wild places, and stare into them and record rituals that the local people perform to bring them into sync with the wildlife around them. There's a man based in Gloucestershire who'd shared a wonderful tale of his time with the people of the Kalahari. He'd gone with them on a hunting expedition and had seen first-hand their ability to sense the presence of game animals 'far beyond sight and hearing'.

The tribesmen had beat drums and danced and gone into a trance state, and their heartbeat and the heartbeat of the animal they'd wanted to hunt had begun to pulse at the same rate, which meant the animal could not detect their presence, so they were able to get very close to it. It's a perfectly natural ability, he said, which the majority of us in the West have no inkling of, no respect for, no interest in – our great loss.

In the afternoon, Elen takes me on a drive through the Golden Valley, which winds its way between the Black Mountains and the River Dore. I try to pinpoint what makes this feel so different from other stretches of countryside I've passed in other counties: maybe it is the heat, which feels voluptuous rather than suffocating; maybe it is the wheat fields, which give off a Van Gogh-ish glow; or maybe it is that luminous blue sky. Whatever it is, being here feels as satiating and as exotic as any wandering I've done on the other side of the world. It's as if I've been let into a secret world of peculiarly English beauty. This feels like a taste of… belonging. I could do with more of this.

We turn down a road into the village of Vowchurch and park under the shade of an oak tree. Across the road is a church, St Bartholomew, that Elen is keen to show me. But first, I follow her down to the banks of the river – a stream, really – and take in the water flowing gently under a pretty bridge. The scene exudes a quiet, midsummer perfection, the sort that grannies and tourists swoon over and makers of murder mysteries covet. Then I turn towards the church across the road and gaze at its striking and faintly sinister-looking timber-framed belfry.

I don't want to go into the church – I want to glue myself to the sun, steep myself in its sweetness. It feels all wrong to be indoors and I'm not a churchy person: the nuances of architecture and history are lost on me and Christianity is not my faith. Whether I'm in a temple like the ones I've visited in India or very occasionally in a church for a friend's wedding or a funeral, it's the atmosphere that I respond to: warmth and incense and colour and glow. In this one I feel a quiet austerity and formality as I step inside. Were this church a human, it'd be an emotionally repressed male – and I don't do well with those.

Elen points out a sheela-na-gig, a naked female figure carved on a wooden pillar. In such figures, she explains, the vulva is traditionally splayed open but here it has been blocked off. There is a Green Man, the ancient spirit of nature, carved on a neighbouring pillar and the lower part of his body has been censored too. Both figures look more weathered than the rest of the woodwork, and Elen says that an earlier pre-Christian chapel likely stood here in the place of the Anglican church we are standing in now. As we are peering at the figures, we hear shuddering in the rafters. It's a swift that has somehow found its way in and is flying around in panicked fashion. Elen tries to coax it to leave but it doesn't work. I'm not sure who is more distressed: the bird or Elen, worrying about both its fate and her inability to connect with it while I stand watching.

To distract her, I suggest we take a look at the churchyard. Outside, she shows me a yew, all squat and split and gnarly and gingery-purple. It is over a thousand years old, which in this neck of the woods isn't *old* old, but still I'm awed. The yew is the sacred tree associated with churchyards and, before that, pagan cultures. I'd read somewhere that in past times, yews were symbolic of long life and that some believed the trees had supernatural beginnings though now people also associate them with poison because the seeds of the yew's red berries are toxic. I remember Charlotte telling me she mistakenly ate some last year and landed herself in A&E, having to have her stomach pumped.

Elen is embarrassed about the swift in the church, I can tell – maybe she feels her inability to personally send it on its way has dented her credibility in my eyes. As we talk a swift now darts back and forth at light speed between us, a blur of wing tip. 'I'm sure it's the bird from the church,' I say hopefully. Elen looks doubtful but points us to a velvety grassy patch and we sit down and rest. With the sun on our faces, now that I'm facing the flowers in bloom and the carefully tended tombstones, there's nowhere else I'd like to be than this warm, friendly-feeling churchyard. But what on earth am I going to do in the Welsh Marches for the next two days if Elen isn't going to teach me the Old Ways?

* * *

The next morning, I find out what we are going to do. After breakfast in the garden, in the midst of what sounds like a glorious bird *a capella*, my host spreads out a big map and points to something called Arthur's Stone, which archaeologists describe as a Neolithic burial chamber. 'We're going here,' she says. I'm relieved we have a plan, but my heart sinks too: standing stones, the ones believed to be the sites of ancient worship, have never moved me. I can never see life or history in them, only inanimate stones. I can't feel any 'energy' in them either, the way some of my friends do. I may be poorer for it, but it's how I'm wired. I've never been to Stonehenge or Avebury, never truly felt the urge to go.

I feel a similar ambivalence when it comes to ancient stories. I might have been born here, but none of the myths or tales of this land are in my bones. The words 'magic', 'sacred' and 'mystical' I can get – to me they feel big and universal and timeless – but I find it hard to connect to tales of King Arthur, Merlin, the Green Man and all the rest. Pick any figure from British mythology and I can tell you almost nothing about him, her or it. So many books on the subject have been written by people who seem to have been raised on the very teat of this land, but try as I might to read such books, to get into them, too often my eyes glaze over. I have never truly felt the pull to explore these… well, I don't even know the right word. Mythical figures? Spirits? Deities? And as defined by whom? People talk about their Celtic connections and myths and stories with so much pride. So are these stories only 'for' people who've lived here in Britain for generations and generations? It's ironic that, despite my love for all things mystical, I have an Achilles heel when it comes to British mythology. I have come to the conclusion that a natural affinity for British folklore just isn't part of my DNA. For example, it has taken me days to make any sense of the scraps of information I have on Elen of the Ways: it has been an exercise in head-banging and frustration. Maybe I feel alienated because, as Elen later says forcefully, 'It's so easy for modern educated folk to fall into the trap of trying to put our Old Ones, our Powers… into ready-made boxes that fit with all the garbage spewed by the average academic.' I have nothing against academics. But I need something, some guidance, for too often I am clutching at straws.

To be fair, I'm not especially interested in the folklore of my Indian ancestry either. My mother would call herself a Hindu – though in reality this means singing Sanskrit prayer songs with a group of her friends, before tea and gossip – and I can barely get my head around the pantheon of deities you could call upon in the religion. She'd light a Hindu god lamp and dab ash on my forehead to bless me from time to time, but my psychologist father was avowedly rational-minded.

In India, I loved the way you could find a shrine on the corner of every street. But ask me what each god represents or what stories are associated with which deity and I'll give you a blank look. I have the same feeble reaction to any mention of the Ramayana and the Mahabharata as I do to British stories. I know snatches of things, but I never feel a sense of belonging to one tradition or any tradition for that matter. Is this the fate of a 'citizen of nowhere'? One day, I promise myself, I will delve deep into my multiple identities, expose them like a rainbow, one vibrant colour at a time.

But now is not the time and I turn back to the matter at hand. I am struggling to understand Elen's connection with Arthur and Merlin, who both seem to have something to do with the Arthur's Stone. But these aren't the Arthur and Merlin of medieval lore – not that I'd understand a whole lot more if they were. 'The stories you're thinking of were written by Christian monks,' Elen says. She's not in the mood to give me her thoughts on a pre-Christian Arthur, but she has written a book on Merlin and in it she describes him variously as a spirit of nature, 'the spirit in its masculine form', and a priest of nature. Is it any wonder I'm confused?

'A priest of nature is very well able to communicate with all the animal, vegetable and mineral kingdoms as well as humans, and this requires that you are able to sense into the essence of other creatures, plants and rocks,' she'd written. Confusingly, at times Merlin had incarnated as a human as well as a spirit. 'He appears in many guises as the Green Man, the Wildman of the forest, as an old wizard, a young boy...' So was a priest of nature a spirit or a human? I feel like I'm sitting an exam, Mythology 101, and failing spectacularly.

In the garden I raise my eyes skywards and think I'd do anything for a good long hike, for this is the kind of summer's day I spend half the year wishing and praying for. There is not a single cloud in the sky and the sun is blazing and it's still just morning. It's set to be a scorcher. Besides, I've all but given up on the prospect of getting in any tuition from Elen in the art of 'kenning' or glimpsing nature as a priest of nature might do. And now I am off to see some Neolithic chamber. Not my idea of a good time.

I'm a bit disoriented on the drive, but again we dip and turn through the Golden Valley and up into the hills and finally we slowly make our way along a ridge till we come to Merbach Hill. Perched in the middle of it are some upright stone slabs on which rests another giant flat stone, though one section of it looks to have fallen away at an angle. 'Very Flintstones,' I think, a little rebelliously.

I simply cannot muster the solemnity and veneration that I suppose I am meant to, given that the smart English Heritage sign here says that the Arthur's Stone is over 5,000 years old (Elen says, no, that it is 10,000 years old) and that the various stones are all that remain of a prehistoric tomb. Rituals may have taken place here, I read. The first time she'd come up here Elen had wondered why the ancestors had built this structure. What could it have been for? Why were so few bones found if it was a tomb? But these are her passions, her concerns – not mine.

The view across the valley, though: this I can worship. The countryside, the patchwork fields, the Black Mountains in the distance, every tree just so. The sheep grazing. It's richly textured. Like a tapestry, a work of art. Or a kind of pastoral sensuality. Lush and benign. The grass sparkles and there is the playful pull of the sun. There are a few clouds overhead but they are harmless, drifting shards of cotton wool. I feel something in my heart yield, and my resistance to being led here falls away. In the distance is a pointy peak. 'Sugar Loaf mountain,' says Elen. 'It's in Wales, across the border.' Beyond it are the Brecon Beacons. This is where the magic lies, I think, when my heart connects to the land. Maybe these stones are like the labyrinth or the treasure map, leading you to an anointed place. A breeze tickles my face, the softest welcome.

'I wish I could camp here,' I say to Elen. It turns out she has. And not just because of the view.

'We're on a ley line,' she says. A ley line, she tells me, is an ancient thoroughfare, part of a network that takes in significant sites – standing stones, church sites and prominent natural features. (At least that's one definition, to some a leyline is a sort of energetic highway). 'This road,' she says, pointing to the track beside the site, 'is an east–west ley line, found by Alfred Watkins' – the man who coined the term 'ley line' – 'but it's also a crossroad with a rough north–west ley line. It runs through the Stone, and points to Sugar Loaf and across the Severn to Dunkery Beacon on Exmoor. This suggests just how cosmopolitan our ancient ancestors were, far more than in medieval times.'

I nod sagely as if it all makes perfect sense to me. But it doesn't: it is word macramé. It doesn't help that I have no sense of direction either. For Elen, however, this is a powerful place. She waves aside the official story – the English Heritage one – that Arthur slayed a giant here. That particular myth has nothing to do with what is going on here. Which begs the question, or at least it does to me, what *is* going on here?

Elen likens the Arthur's Stone, the prehistoric chamber, to a womb. The first time she visited it, she crawled in and had a vision. In it she saw Ceridwen, a three-faced goddess who is 'the maiden, the mother and the crone,' she says. 'Ceridwen can offer inspiration, nurturing and wisdom.'

I am even more befuddled than ever now. That is the problem with folklore – the stories seem to rise up from a murky cauldron, like steam, will-o'-the-wisps that you can't get a handle on. The stories are like spirits themselves. But the point of telling me all of this, of bringing me up here, my host says, is that any one of us can crawl back into the chamber or the 'womb' and 'listen there for the Old Ones to speak and show us things, and then crawl back into everyday life.' A kind of ritual then, to connect with the Other, can be performed here.

'You want me to crawl into the chamber?' I say, half-fascinated, half-horrified by the idea. Elen nods.

So that's what I do. With a glance left and right, just to be sure the coast is clear, I drop to all fours and crawl along, praying that no hiker will come by. I ought to be hiking myself, not doing this, I think. I don't wish to be impolite though, and besides I am genuinely curious: can I, as Elen suggests, connect with the Old Ones who come here? Might 'they' have a message for me? But which Old Ones or One? Ceridwen? Arthur? Merlin? I feel like I once did in my ninth-grade physics class, totally and utterly mad with incomprehension. But there is the slim possibility that something might actually happen and I am holding out for that.

Into the stone 'entrance' I inch, trying not to graze my knees. I manoeuvre myself onto the cool, restful slab and attempt not to bump my head. At least I am out of the burning sun. I whisper 'show me what I need to know,' as Elen has suggested, and I feel ridiculous saying this because I don't know who I am asking to show me whatever it is I need to know. My rational brain is thrashing around, looking for something to wrap itself around, and then I think: 'Oh, hang it,' and I let go and begin to daydream.

★ ★ ★

Half an hour later, back out in the sunshine, I relay my daydream to Elen. It turns out I have a pretty vivid imagination. Glittering red stones are involved. Lucy in the Sky with Diamonds, only with rubies and a wizard. For I had conjured my very own Merlin. 'You're not the only one to have connected with him up here,' Elen says mysteriously. 'You're in a powerful place, which means you're more open than usual.'

It's for me to figure out what the message means, she says. So what I do think? Well, the experience, though pleasant enough, feels like nothing more and nothing less than a deliberate daydream, a fantasy – not even a proper falling-asleep dream. I'm disappointed. For a long moment we don't speak and just gaze at the views. 'You can walk all the way down the valley from here to the River Wye,' says my host. And with every fibre of my body I long to do this. But today it won't happen.

Instead, we drive to the village of Peterchurch, where we buy an enormous tub of vanilla ice cream from a shop where the only other customer is speaking to the cashier in Welsh, and we take it to the grounds of the church. The towering trees in St Peter's would not look out of place in a nature reserve or a National Trust garden. On the north side of the church, among the tombstones is a 3,000-year-old yew. *Three thousand years old*. That makes it extremely rare. I later read a local newspaper story about it, which tells me this yew is one of the 12 most ancient trees in the UK. I have a look the church website too and find a whole section on yew mythology:

> *To the ancient Celts they were called the Tree of Life and they have been viewed as sacred and magical across many cultures. In Celtic paganism, they were considered a portal to ancestors and a means of communicating with the dead and they were therefore planted on sacred sites. A common, more recent explanation [for] the presence of yew trees in churchyards was that they were planted for wood which makes excellent long bows, but this theory has now been abandoned as the volume of yew wood needed for war archery during the Middle Ages was far too great to be in proportion to the wood which could have been grown in churchyards. It is now known that the trees were there long before the churches and are the result of the early Christians adopting certain aspects of the pre-existing Pagan beliefs into their own, so that churches developed naturally on pre-Christian sites where yews have long been revered.*

I'm impressed to find all this information on a church website. This tree is so wide it'd probably take four humans to wrap themselves around it. One side is cracked open, wide enough for a child or even an adult to climb inside, so with Elen's encouragement I do. 'Hello, yew,' I say. A being that has survived this long deserves respect, like a wise great-grandmother. It has an aura of antiquity, not unlike something I might find in the V&A. I crawl back out and we sit in its shade and cram spoonfuls of the ice cream down our parched throats. In our line of sight is a towering redwood, which Elen reckons is around 150 years old. A mere stripling. 'It's the yew tree's partner,' she says.

Maybe it's the kind of day we've had, or the frame of mind I am in, but that doesn't sound ridiculous to me: the trees *do* look well matched. After a while, we leave the churchyard and walk back to the car park. As we do a kite flies towards us and circles overhead. Elen points out the distinct rust red of its forked tail. I've never seen a red kite before. It wheels around, riding the wind, its wingtips ever so sensitive. Elen becomes jubilant, all trace of tightness in her voice gone: the transformation is astonishing. To her the bird is a sign, a deliberate wink from nature. It is everything our encounter with the swift in the church yesterday wasn't, and I can see she feels redeemed somehow.

★ ★ ★

I decide to return to London a day early – I think both of us are relieved by this change of plan – and on the morning of my departure, I am packed and ready to go pretty early. I sit in Elen's living room and admire her collection of handmade drums while I wait for her to come down. The weather is still golden and I think that if it looks like this back in London, I'll go for a walk in the woods. I'm glad I've booked an early train. Maybe I ought to have booked an even earlier one. Suddenly Elen rushes downstairs in her nightgown and thrusts her laptop in my face: 'I think you need to see this,' she says excitedly. I've not seen her this excited before, not even when we saw the kite.

I read out the email on the screen. It appears to have been sent by a woman in the West Indies the night before. She'd had an odd dream that made no sense to her and had sought an expert to help her interpret it. Randomly, she found Elen online. In her dream, says the woman, she meets Merlin and he tells her that if she needed healing, a ball of fire would appear in his hand, and he would hold it inside her head or directly to her heart. I shake my head in disbelief. The woman's dream is spookily similar to the daydream I had up on the Arthur's Stone. How is this even possible? Is this email legit? I look at it closely. Yes, it's legit. I don't know what to make of this. The dream isn't some kind of common lore that you can easily look up and regurgitate. This is a real person's experience, as real as my own. I look up at Elen, and for the first time since I've arrived our eyes meet properly.

Hurriedly, I rummage around in my rucksack for my notebook and flick back to the words I'd scribbled on Merbach Hill the day before after my time in the chamber:

> *Merlin leads me by the hand into a cave where we sit round a table. He points to a spot in the middle of my forehead, and then to a red jewel on the table in front of us. He gestures for me to put the jewel in my heart and then in the middle of my forehead. My true power comes from my heart is what he is saying. I nod and in that instant he morphs into a woman with long brown hair. 'We're both one and the same, the spirit of the land,' she says. At this point my conscious mind is relieved that I have managed to come up with a daydream at all.*

The synchronicity is quite something, and to me not a fluke or a banal coincidence but an indication of some kind of elemental connection being made. That it makes no sense logically and will live in the box marked 'mystery' doesn't make it any less authentic.

'You needed proof that you really did connect with the spirit of the land. And the spirit of the land found a way to reach you,' says Elen. 'There are no coincidences.'

I've set out to explore the land in search of the magical, but what I hadn't anticipated is that the magical might reach out to me too. So now I am in uncharted territory. Life is stranger than fiction and life is about to get stranger still. My head is whirring.

# The Secret Place of the Wild Strawberries – Part I

I need a time-out, a break, a blast of fresh air, a long and uncomplicated hike, the sea, nothing more. I don't want company. Everything has become too intense, and claustrophobia is rushing in on me. I've decided to take the train up to Berwick-upon-Tweed and from there head to Lindisfarne. I have a special place in my heart for islands. I was born on one and I grew up on one, so it makes sense.

First I have to endure the noisy people on the train. The woman in the seat opposite rolls her eyes and we bond in mutual annoyance at the four raucous flight attendants at the table seat in front, who are still in uniform, and swigging gin on a Tuesday afternoon. The delicious, wild, unspoilt beauty of the Northumberland coast, those miles of white beaches – the closest thing to Namibia's Skeleton Coast that I've come across in Britain, seals included – that's still to come. When the seascape finally does appear, it changes the atmosphere in the train, lightens it, so that I feel like I'm on one of those levitating Shinkansen bullet trains, peering down at the shoreline. Somewhere out beyond, swimming in the North Sea waters, are dolphins and even whales.

At the station, I climb into a cab and off we race, against the incoming tide. One of the great joys of the island is that when the causeway closes, no one can get on or off and a strange intimacy descends. Humans and wildlife and birdlife, all together on a little ark.

I'm here because I want to get away from people, clear my head and explore the empty beaches and the vast silence of the island's north coast, as well as all those neighbouring coves and rocky headlands and beaches. I want to know if I can hear anything in those silences. I want to know if I can make sense of some of what's been going on, or alternatively forget all about it for a bit. Leaven the mix, gain a bit of perspective.

Many centuries ago monks were sent here from Iona, the Hebridean island up in Scotland, by Aidan, himself a missionary monk who'd been summoned to Lindisfarne by a king, Oswald. Aidan needed the monks' help to establish a monastery and convert the non-believers to Christianity. When he died Cuthbert took over. History describes him as a nature lover, a shepherd, pious yet humble, and a kind man who performed miracles and craved solitude. Most people who come to Lindisfarne today make a pilgrimage to the Priory and to St Cuthbert's, named after the monk. The little islet cut off from the main island at high tide was where he went on retreat. Both of these places are on the south side of Lindisfarne. Most of the day trippers ignore the isolated, wilder, north coast in their eagerness to pay their respects to a monk who shied away from people, which is a little ironic but also good news for me.

I'm staying at a Christian retreat house here because it's all I can afford and payment is by donation but I genuinely do relish the idea of a few days' semi-seclusion. It's not long enough, I know, but it's better than nothing. I just hope the cost won't be a bible-thumping.

The house is a great big Georgian one in the centre of the village, on a quiet cul-de-sac. It has a beautiful garden, which will fill with birdsong in the morning, and a library – a silent room – and a sitting room and even a kitchen in which some thoughtful person has put out a tray of home-made brownies with a sign that says 'help yourself'. I would, but the warden is with me, showing me around and I feel I need to be on my best, most solemn behaviour to compensate for being slightly heathen and visibly non-Christian. On its website, the retreat centre says it is not meant for holidays – I think they mean the sort that involves lots of booze and late nights – but for me four days of reflection and walking *is* a holiday.

I've arrived early evening and dinner is across the road in another house. It's a communal meal, which I don't mind because there is no pressure to socialise at any other time of the day so I'll be otherwise as free as a bird. I am hoping and praying, in my secular way, that the conversation will not be too theology-focused because then I will be out of my depth. I imagine I'll dine with some elderly cleric or retired schoolteacher or a church-going single gentleman birder or whoever else I presume comes to stay in a place like this.

The meal throws up the first surprise. My fellow retreaters are all women and all very different and all very interesting. They are not the sort of people I might have expected to meet in such a place, except maybe Raewynne, who is a cheerful and funny Anglican priest – 'a priest of God as opposed to a priestess of the Goddess,' as she puts it. She is originally from Australia and her kindness is matched by an enthusiasm for all things Christianity. 'This is my space, ask me anything you like,' she says, beaming.

I'm tempted to ask her about celibacy and dating if you're a priest and how it all works – the first thoughts that pop into my head uncensored – but wisely I hold fire. As I understand it, she has hiked all – or some of – St Cuthbert's Way, a walking trail from the Scottish borders to Lindisfarne, named after the hermit monk. Raewynne is a true pilgrim and did the last leg (the causeway over to the island) barefoot.

Mary is a Tibetan Buddhist who lives quietly in Yorkshire. 'I'm originally from Northumberland, though, so I am of this land,' she says with pride (but not in a jingoistic or overbearing way). She cares about environmental issues so we're on the same page there. Her spiritual leader has been disgraced by a scandal and she is debating whether to return to Christianity or not. This week will help her to decide.

The final woman, Maria, is from Argentina and was born into a wealthy family. She'd worked in construction her whole life but then developed all kinds of chronic health problems. This led her on an odyssey around the world in search of healing. She has been to see the controversial guru John of God, known for his gruesome-sounding psychic surgery procedures and his miracle cures. Now she has come to Britain. 'I'm on a tour, visiting your most sacred places,' she says. 'We're doing Goddess worship in Glastonbury, then we're on to Avebury, Stonehenge and Tintagel. I'm here because I want to visit the Priory.'

And then there's me. What to say? I tell them I am the daughter of an atheist or agnostic father – I'm not exactly sure which he was – and a Hindu mother and that I went to a Protestant school and that nature is my God. All in all, collectively we cover a lot of bases.

For the women, this week is about getting closer to the Divine, within the parameters of organised religion – at least that's how it seems to me. None strikes me as being especially mad keen on the outdoors, with the exception of Mary, who lives a rural life. Over our communal dinner in a pretty book-lined library – this place is not short on books – the conversation turns to saints and biblical references that mean little to me. My mind wanders over to the shelves and, amid all the Christian spiritual titles, one catches my eye: *The Divine in the Landscape*. Aha!

The title sends a shiver up my spine. Is my search for the magical, the Other in the landscape, the elemental life force in nature, a metaphor for a search for the Divine? In any case, seeing those words gives me courage to join the conversation. I turn to the women: 'For me the Divine is in the landscape,' I say authoritatively. I feel good now. But the women merely nod and carry on talking about the saints, a female saint in particular, Julian of Norwich. I had no idea Julian of Norwich was a woman. I didn't know she wrote a book called *The Revelations of Divine Love*, that she had visions. Everyone in the room is a Julian of Norwich expert, and the warden joins in the conversation excitedly as he collects our dinner dishes. I feel like everyone is speaking a foreign language. The conversation is too tight, too dense for me. I need air, literally and figuratively. That's why I've come to this island, I remind myself. Not to shut myself up in a library, in my head.

I excuse myself and head off to watch the sun set, and to get my bearings a little. The quiet that surrounds me as I walk down to the causeway is a mile deep, pure and soothing. I need this. I inhale, hold my arms wide. The air is crisp, a little sharp, the kind of air that cuts through your thoughts and sets you right.

I can see the pilgrim posts visible in the water: here devotees like to recreate the rituals of the past by walking across the causeway from the mainland when the tide is low. It takes time and commitment, and the sand and water are cold underfoot. I find a pretty bench and park myself on it to wait for the sun to slide into the sea, hoping the air – fresh and sharp as a whip – will help me shake off my tiredness. The light begins to fade, just a little, and the sky begins to seep orange streaks. I am diving into the peace, finding my place in it, finding my way back to myself and then a couple walk on to the beach. Hand in hand, they face the water and continue the thread of a conversation.

I stare hard at their backs, willing them to hush.

'Enjoying a bit of meditation, eh?' says the man when he notices me.

I smile politely and it seems this is invitation enough for him to come and sit beside me and talk at me. I try to split my head in two – eyes desperately fixed on the horizon, ears politely paying attention to what he is saying. Meanwhile his wife stands quietly beneath us on the rocks, seeming grateful for the reprieve. His sharing comes at me in staccato bursts: his favourite spot on the island is the Crown and Anchor pub. If I want to I can walk round the island in two hours. And there's nothing much worth seeing beyond the Priory. 'You want wildlife? Go to the Farne islands, at least there you'll spot the puffins,' he barks.

He rambles on, the very picture of privilege. The kind of privilege that gives you carte blanche to barge in and foist yourself on another. I've been talked at by men like this my entire life. 'People don't come here for nature, they come to make pilgrimages,' he says with a dismissive shake of his head. He should know, he is originally from Northumberland, now lives in France, and has come back on a sort of nostalgia trip. 'I don't care much for what the Archbishop of Canterbury has to say. I'm more inclined to listen to the Archbishop of York,' he says. On and on he goes, his wife ignoring us and enjoying the sunset that I too am trying to savour. Why *do* I listen politely? I didn't come here for this.

'I'm going now,' I say stiffly and get up and walk away, back to my guesthouse.

I'll have to hope for another sunset tomorrow. I want to be hypnotised by the dunes, lose myself in the marram grass, feel the love pour out of me.

⋆ ⋆ ⋆

'But you can't really conjure up things at will like that,' says Mary the next morning.

We're sitting in the glorious garden, surrounded by birdsong. The island is a renowned spot for bird-watchers, people who will have a keen eye for the migrant birds – unlike me, on first-name terms only with the blackbirds and the robins and sparrows and the magpies. We're eating bowls of porridge, sipping tea, and have ascertained that neither of us is in a 'silent' space. There is a certain etiquette to be followed. If you don't want to talk to people you avoid eye contact, make a point of not smiling. We'd emerged onto the lawn, both of us smiling, like a pair of matching cards in a deck, so it's OK. I've told Mary what I'm up to. She wonders if deliberately seeking 'something more' might work against me. Yet it has become my new normal, my new bottom line, I think. And that thought excites me a little.

The sun is bright and white, its rays urging me on. It's a perfect, clear day. Unfortunately for me, I've overslept and it's past ten. 'You need to rise earlier,' says Mary. She herself was up at six, has been to the harbour, watched the sun rise, attended morning prayers. 'And don't try so hard. Good luck.'

With Mary's words ringing in my ears, I stuff my packed lunch in my rucksack – one of the helpers in the kitchen at the retreat centre has made me a sandwich – as I try and make a plan for the day. I'm feeling lethargic, which doesn't help. Should I take a footpath near the causeway to the dunes and then cut across to the north shore? 'Beware the patches of quicksand,' the warden said last night, which had scared me a bit. Or should I strike out past the harbour, towards the castle, currently half-hidden in scaffolding, and then onward to Emmanuel Head and to a rather mysterious-looking pyramid? From there I could explore the beaches and coves, until the North Shore is within my sights. And then play it by ear. Yes, I'll do that. I fold up my Dinky Map and finally get going.

I pass a hipster coffee house – amazing that such a thing exists here – a National Trust shop, an intriguing entrance to something called the Scriptorium, and a pub outside which four Japanese tourists are loading up a car. At the end of the street I've walked down, a path opens up and the village peters out. The path has a 'Yellow Brick Road' feel to it. It's wide, there are fields on one side and the sea on the other. And there is no one else on it. Literally no one. I can see the scaffolding-clad castle in the distance, and the harbour nearby with its old overturned whale boats. I can hear too the comical call of the birds that Mary will later tell me are eider ducks – they sound like a bunch of gossipy humans exclaiming loudly. I decide to save the harbour for the following morning's sunrise.

Before I can find my rhythm and begin to walk myself into the island, I'm distracted by a sort of concrete passage. In it is a wildlife display. The information panels tell me that in winter and spring birds flock to the island to feed. Pied wagtails, meadow pipits and black-headed gulls, arctic terns, all of them escaping the cold. I've seen this exhibition once before: two years ago I came to the island on a November afternoon. The weather had been frigid, the twilight a dusky-pink showstopper. I'd come with Martin, a wildlife guide, for a night of stargazing – Northumberland has some of the darkest skies in the country – but first he'd driven me to the Fenham Flats on the mainland, part of the Lindisfarne Nature Reserve. This was a winter home for birds escaping the cold in Iceland and Scandinavia. It was serene and windswept but oh so melancholic.

I'd done my best to share his enthusiasm, to spot the birds through the binoculars he'd lent me. 'Pale-bellied brent geese,' he'd said, pointing to the birds, a blur of black and brown and grey, barely visible on the mudflats. In the half-light, I'd made out the Priory ruins and the castle: they looked ghostly. We'd got back in the car and driven over to the island. It had been pitch black by then – the night had fallen so fast, and the stars were just visible by the time we'd walked over to the wildlife exhibition. I'd had no sense of direction, of where we were. But the idyllic pictures of the North Shore and the dunes had caught my attention much more than the birds in the pictures.

'I'll come back next summer,' I'd promised myself. I had wanted to walk among those coastal dunes, far from the priory and the village. They'd felt alluringly, deliciously wild, devoid of human footprints, leading somewhere I was sure I'd want to go. Then we'd gone back outside, Martin had set up his telescope and we'd gazed at the Milky Way, visible thanks to an absence of light pollution. The light show was astonishing, the belt of speckled gold in the sky reminding me of the unfathomable vastness of the universe.

'The Milky Way is just one galaxy among billions that make up the known universe,' Martin had said. How could any human every come to terms with the cosmos, I'd wondered? Ever be anything but humble in the face of mystery that is life? The Incas revered the stars, saw meaning in them. They believed everything in our world was connected. As did the Kogi who live deep in Colombia's Sierra Nevada de Santa Marta mountains and who believe in a cosmic consciousness that is the source of all life and the 'mind' inside nature too.

When I think about this, I feel a deep gladness that I've returned to this beautiful island, and in springtime too, bursting as it is with life. I leave the concrete and the bird-life display and go back out into the sunshine. I can feel the heat rising already and stop to unpeel a few layers. Fortunately the causeway will not open for a few more hours, so peace reigns – a gift of the tides. I follow a footpath around the castle and stop to chat to one of the workers. 'Do you know that seals like to bask over there?' he says, pointing to a sandbar. I shake my head and wish I'd thought to bring binoculars.

Beyond the track, close to the shore are strange little towers made of stacks of fat, heavyweight, smooth stones layered by human hands – maybe someone on the island has been practising the art of stone balancing. There are so many of these, like bird droppings, that I almost miss what is front of me: a labyrinth! How unexpected. It seems I can't get away from them.

I take off my shoes and socks and walk it. I can hear the lapping of the waves, a gull overhead. What is here for me on Lindisfarne, I ask? As I walk round and round, sunlight dazzles me with every circuit. By the time I reach the centre of the labyrinth, a glimmer of an answer has come to me. The oracle has – finally – spoken. Like a cosmic road bump, the labyrinth has stopped me in my tracks. I'm here to remember the power of tangible, physical nature, staring me in the face. The spirit of nature, alive in the salty, icy sea, the clammy seaweed on the shore, the rocks, the marram grass, the birds in flight. Sometimes what is here in front of us is all we need.

I can see Mary out of the corner of my eye, walking slowly towards the shore, so I decide to leave the labyrinth to her and continue on the footpath, now parallel to the sea. It's all so easy, so unrushed, so alluring, so sweet and peaceful with just the lightest breeze, like a distillation of every good feeling I've ever craved. So why do I not do more of this? Because I live in London, because the tentacles of the internet have me in their grip. Because, despite my best intentions, I spend more time scrolling than walking and feeling. Because I'm broke. But here the cloud has lifted and I can see and feel the truth: life is *this*. This is how it is meant to be.

The morning slides into early afternoon, and everything is a blur of walking, white sands, rocks, wind and a mournful wailing. I stop and listen. Who is doing the wailing, I wonder? Is it the wind? Is it the seals?

Disappointingly, the striking Colgate-white pyramid at Emmanuel Head is not some mysterious ritual object as I'd hoped, but a navigational aid for boats built in the 19th century. Still, it makes a good lunch stop and I sit on a bench and eat my sandwich as the wind whips my hair up and around my face. It's so violent I wonder if the Furies have swept past, and I think of all the vessels that have crashed on the rocks, all the sailors who have lost their lives. When I reach Sandon Bay I stand atop the dunes, unsure whether to go down to the beach. If I do, I might not make it round the island. I sigh, impatient with my petty anxieties, and shuffle through the sand, navigating pebbles, till I reach the beach.

I notice a couple, barefoot and perched on the rocks in a corner of the beach, sitting in a lazily proprietorial way. I tactfully choose another spot a little further on, then dump my rucksack, unlace my boots, peel off my socks and walk gingerly to the water's edge. All these tiny actions feel like a commitment. Which is crazy. It's only a beach. But it takes *energy* to do this. I am a woman who has taken herself off to Northumberland and the tiny island of Lindisfarne to walk alone and here I am. No big deal. And yet at the back of my mind are all the headlines shouting about people whose skin colour is not white who never connect with nature. It annoys me, the generalisation. I'm just a human enjoying the sea, like any other. I've been doing it for a long, long time. Like other women and men before me whose storylines don't fit the plot and whose voices are never heard.

The water is icy cold. I do a little hopscotch on the edge of the water, just because. *Splash, splash, splash!* And then I hear the moaning again, carried on the wind and stopping me in my tracks. Is it a trick of the sea? I can't see any seals.

I could spend a whole day hanging out on this beach, I think, if I already knew the lie of the land. If I had more time. There are more micro-decisions to be made. Do I carry on walking till end of the beach and then climb back up to the biscuit-coloured grass and the dunes and carry on to Castlehead Rocks and Coves Haven from there? Or do I retrace my steps and climb up to the sandy footpath, and then continue on?

I can't work out why I am finding it so hard to make these tiny decisions. It's not like I have to worry about storm clouds or rain – the sky is a clear, cornflower blue.

Can't I just sit on the pale sand in this untroubled bay and enjoy the waves rolling in and out? 'Slow down,' I hear the waves whispering to me. 'Slow down. These are my secrets.' The secrets are encoded in the lapping of the water and can only be known via feet in icy water. The mysterious wailing stops, and eventually I rise and retrace my steps. And then I get lost.

I'd imagined it would be a simple thing to carry on over the dunes to Coves Haven, beyond Castlehead Rocks. But I hadn't reckoned on the land playing tricks on me. I veer inland a bit as I don't want to end up on Keel Head, a thin, almost phallus-shaped headland jutting out at the end of Sandon Bay. But when I lose sight of the coast, I begin to feel disoriented, like I'm in some sort of no man's land.

I'm surrounded by grassy dunes, the long fronds waving in the breeze, taunting me. This could be some desolate American wilderness. This can't possibly be England, be Northumberland, be Lindisfarne. I trudge this way and that and begin to wilt in the heat: not a thing one often says in Britain (but then again we are in the midst of a climate crisis). I may as well be crossing the Thar Desert, I think, eyeing up the paltry half-bottle of water I have left. Of course, this isn't a desert, and the scrub is scrubbier there – nothing like this at all. But I don't care about the detail; it's the impression that matters to me and I am feeling that kind of deserty desolation. Right now the dunes are the boss of me. I want to find my way back to the coast and Castlehead Rocks, but which way is that? Suddenly I remember my phone's GPS. I have completely forgotten that it also has a compass, otherwise I'd have used that, the less offensive of the two gadgets. I hold the device covertly, in the palm of my hand, praying that no one comes along and catches me at it because then I morph into a stereotype, the 'urban ethnic' woman unable to part with her phone, even out here on Lindisfarne's wild north coast.

I'm not too far off course, as it happens (small congratulatory pat on the back). I turn a corner and there in the distance is a blunt, flat slice of rock. I think this might be Castlehead Rocks. From here it looks like a mini Uluru in a not-quite-Outback.

I wonder if the monks who lived in the Priory ever ventured to this blustery corner of the island. The rocks hug the cove to the east, and the waters here are deeper, darker, a velvety purple. There is more tangled seaweed down on the shore too. Everything about this beach feels a little more dangerous than the first, as if I am crossing a threshold. Even the sky is cobalt here –the kind of blue that always taunts me when I'm alone in a beautiful place. A blue for gazing upon with a lover. The wailing is back but it's all knotted up in the wind, so it's hard to tell one apart from the other. Is it the pod of mammals perched far out on the rocks, bobbing up and down? I once spotted a colony of seals on the Skeleton Coast in Namibia, and I'd been stunned to see them in their hundreds, an irascible rabble arising from nowhere, beyond miles and miles of empty, pristine, white sands.

I sit here for ages, boots off, socks off, feet on the hard sand, trying to memorise the peace, the red of the rocks that I find reassuringly exotic and therefore familiar. I need all of this to penetrate deep so that I can carry it around with me when I leave here. Seabirds I can't name reel about in the sky above, and a strange orange furred caterpillar creeps towards me then politely changes direction.

I lace up, leave this beach and carry on over the dunes towards the North Shore sands, walking past a pile of plastic bottles and then another dump of rubbish, all of it a sign to not go further, to not venture onto the maybe-quicksand-maybe-not that I can see stretching out so tantalisingly close. But I'm tired and so I turn back, south across the dunes, following this path and that, not getting lost this time, my eyes glued to the pilgrim posts that jut out in the mudflats, the last leg of St Cuthbert's Way, which occasionally traps drivers who ignore the tide tables and the warnings.

* * *

I've decided to go to church. St Mary's, the parish church, is in a corner of the village next to the ancient Priory. It too is old. Very old, in fact, well over a thousand years old and entwined in the history of St Cuthbert and the monastery. When Raewynne, the Anglican priest, sees me poke my head through the door for the evening service, she grins. I've braved it. I am here half out of politeness, half-wondering if I'll feel anything of the sacred in the service. I'm a little anxious, as if I'm about to step out onto a stage, without my lines. At the retreat house, church attendance is encouraged, but I'm not a churchgoer – and why would I be, being the progeny of a non-practising Hindu and an agnostic-or-atheist father?

I'd actually scoped the joint, in a manner of speaking, beforehand when I discovered a website where a 'mystery worshipper' goes around the country reviewing churches. Someone with the pseudonym 'Kettle' does the honours at St Mary's. 'Was the worship stiff-upper-lip, happy-clappy, or what?' is the first question posed. Kettle replies: 'Wishy-washy, almost traditional, but not going quite the whole way. Quite reserved'.

It's the 'reserved' bit that puts me on edge. What exactly does 'reserved' mean? Icy? Suggestive of an intolerance towards non-Christians? I read on. 'On a scale of 1–10, how good was the preacher?' The mystery worshipper replies: '5: She spoke very softly, and sounded a bit bored with her own sermon.' I scan the next few questions, and then sit up straight:

'Which part of the service was like being in heaven?'

'The silence. After a hectic weekend, it was heaven indeed.'

'And which part was like being in… er… the other place?'

I don't think I'll share the reply to *that* one here. The silence, though, that sounds good. Beyond that, I'm clueless. I'd moaned to Raewynne back at the retreat house: 'I won't know what to do!' I've been to churches for weddings and funerals. I've escaped into them when travelling in India, knowing I'll find quiet and peace, an escape from the madding crowd. But I don't understand the formal language used in services. I've not had any kind of religious upbringing, so it is not surprising I don't have a clue about the etiquette of church-going.

But I also don't know why I am quite so anxious about it. Maybe it's because I know I'll be hyper-visible, standing out like a sore thumb. I can't stop the questions that pour out of me. When am I meant to stand? Where am I meant to sit? I have no singing voice, either. Will I be expected to get up and sing or recite a verse from the bible?

Now, when I join Raewynne, she tries to reassure me. 'No one will make you stand up alone and say or sing anything. People will sit across from each other in two pews,' she says, pointing towards the benches at the back. There'll be some kind of call and response. She shows me a board on which I can pin up a note: 'One is for prayers, the other for giving thanks.' She beams – she knows I'll like that bit. The priest will read them out at the end of the service. I spend a few moments thinking about what I want to give thanks for: the insane beauty of the island, the wild, welcoming spirit of Lindisfarne too.

There's a bit of time still, so I read notes on a board about the history of the church, of the island, and about the Lindisfarne Gospels: St Mary's is the oldest building on the island, and stands on the site of the original monastery founded by St Aidan, back in the seventh century. In the gloom, I inspect a striking sculpture of six hooded monks carrying the coffin of St Cuthbert. Their expressions are pained. The sculpture, crafted in elm by an artist named Fenwick Lawson, is called 'The Journey' and it depicts the monks fleeing Lindisfarne with Cuthbert's remains following an invasion – theirs was a pilgrimage to Durham.

I try hard to concentrate and be interested and absorb all of the very important historical information presented here – I've not even mentioned the Lindisfarne Gospels, mainly because I can't quite get my head round what they are other than some kind of treasured book full of biblical references. Any mentions of them that I come across seem to have been written for people who already know what they are, which isn't terribly helpful. The most succinct definition I find is in a passenger-ferry guide, of all places: 'An eighth-century manuscript of the biblical gospels in Latin that feature a stunning collection of illustrations.' So there you go – the Lindisfarne Gospels for Dummies.

My mind is stubbornly resistant, all of this history feeling to me like a hard kernel, distant and non-native to me. I feel the energy of elderly men and their spidery thoughts and their practice of conversion, of persecution, of conquest of those who did not share their beliefs. It makes me shudder for there were among the persecuted, those who worshipped nature and saw the Divine in the earth and the cosmos. What happened to the indigenous people in far-off lands happened here too.

I also have no appreciation for the finer points of church architecture and I need a dictionary to translate words like 'nave' and 'chancel'. I prefer the wild places and the footpaths around the island that are open, and heady and seductive and sweetly inviting. But the light streaming into the multi-coloured stained-glass windows is warming. And the flickering flames atop the candles are here for me and I am grateful for them.

After the blessedly short service with its reciting of prayers and singing of hymns, which I manage easily enough, the prayers and the expressions of gratitude are read out. I'm shocked that in this church – of all places – I'm the only one who has offered a thank you, on this island which gives and gives and where no superlative is adequate.

As we file out – Raewynne smiling at me in approval for having stuck it out – the priest shakes everyone's hand but clutches mine in both of his. 'Thank you for coming,' he says in a heartfelt way that suggests he is deeply thankful. Thank you for coming even though you are Other.

* * *

'I'd had enough of academia,' says Mary Gunn, a local guide and ex-scientist, by way of explaining why she married a lobster fisherman. We've met near the causeway and we're walking to the North Shore. I'd found her details pinned to the window of a village souvenir shop and I figured it might be nice to chat to a local. Besides, she knows all the shortcuts through the dunes. We pass Green Shiel, a collection of stone ruins en route. 'They were probably the homes of the workers who served the monks,' Mary says.

The flowers that carpet the dunes are stunning. 'Viola, lady's smock, dog violets, forget-me-nots…' Mary reels off the names, pointing this way and that. Nature's palette, the comforting blues and sharp purple and pale lavenders are all working on me. We reach the beach in what seems like minutes. 'Quicksand?' says Mary, eyebrows raised, when I mention what the warden has told me. There are a few patches but they are well along the three-mile strip of sand. There's nothing to worry about, she says. Nothing to worry about but the gusting, shifting winds eager to hurl us about like kites.

The tide is out and the sands, mud-coloured under the grey skies, stretch on as far as the eye can see. It's melancholy, raw and windswept here. The grassy dunes that flank the sands are bent over in the wind as if imprisoned, their rustle barely audible. In front of us the sea is agitated and the seabirds are moaning. This all feels too big, too vast, too despotic to be part of a tiny island in Northumberland.

Mary is talking above the wind, telling me there is a ley line between here and Iona in Scotland. 'Ley lines are energetic lines connecting sacred places,' she says. Vessels for a powerful energy. Her definition is the more popular, esoteric version. I'm surprised to hear an ex-scientist speak of such things. 'The monks chose to live on Holy Island and on Iona because of the energy they found in these places,' she says. But what is it that elevates these islands above others? Is it a presence that the monks could sense? Something that can be heard when one sits in silence? Today, this island is giving nothing away.

We walk over to the salt marshes for a look at Snook House and Snook Tower, two old buildings. The house has a lived-in look. 'That's because it is,' says Mary, telling me a story about the family who are renting it. Wild strawberries grow round here in season, she says, but it's too early in the year so there is not a splash of red anywhere. I pay little attention to the grass underfoot, but if I had I'd be astonished by the variety of life here.

We plod back over the muddy causeway and say our goodbyes. I'm leaving the next morning, with Lindisfarne lodged in my heart. But I plan to return when the strawberries are out. I have a feeling there is something more waiting for me here.

## The Tree Whisperer

Back home, it doesn't escape my notice that much as I have an instinctive love of the coast, my passion for trees is also beginning to intensify. Maybe it's a knock-on effect of the walking I have been doing in the woods at the end of my street. I admire the beauty of trees, their slowness, their steadfast charm and their companionship. I marvel at their fruit, their flowers and their medicine. It dawns on me, like a slow-burn infatuation, that I am a tree hugger and so I do begin to hug them, but only when no one is looking. To my astonishment, a tree app, a tree reference book and a tree identification flip chart that fits in the palm of my hand have become my new accessories. It's so unlike me, although as an American friend once said to me in an Oxford college garden filled with beautiful old trees: 'It's a bit rude, don't you think, that we don't know their names?' Put like that, maybe it's time I learn them. But I am also watching myself do and think these things – standing on the outside and looking in on myself. I don't think I am about to make a Damascene conversion and become a lifelong, crusading cataloguer. Though this new hunger to know the names of trees does help me to see that those who are might be fuelled by love too, and so I feel a little less inclined to judge them.

Maybe in a year or two, while my love for trees will endure, this particularly intense curiosity, this ardour may mellow a little (as with any love affair). I am pretty sure I will not become a *taxonomist* – the word sounds hard-edged and alien on my tongue – or a connoisseur of classification. It is not in my DNA. Yet, for now, I can't walk two feet without staring intently at leaves and berries and bark. I take to carrying a paper bag for all the fruit I can't resist snaffling. I consider travelling around Wales in search of ancient yews, but am thwarted because I have no car and can't figure out how to do it over a weekend on public transport.

I rack my brains and then forget about this idea of a tree mission for a bit – it's not difficult, I'm easily swayed and life is full of distractions – but eventually I decide to travel up to Derbyshire to meet Glennie Kindred. She writes books celebrating the cycles of nature, and she's a celebrant herself. She draws, gardens and lives by the seasons, she makes herbal potions and cures, and a few hundred years ago she'd likely have been burned at the stake. Glennie is a font of arboreal wisdom too: she knows the lore connected to all the British native trees, spends time with trees year round, and believes it is possible to enter into a more subtle relationship with them, and to 'talk' to them even, which is the bit that excites me. Glennie, you see, is a tree whisperer. *A tree whisperer!* She has promised to take me to meet some of her local trees and I'm sincerely hoping her idea of 'meet' will involve communing with the spirits of the 'standing ones' (to quote one of her fans). I may as well be upfront about it.

I'd booked the train ticket, but a few days before I'd had one of those weekends where you do one too many good things. I'd spent the Friday with an environmentalist friend who works for DEFRA but who on the side secretly bangs gongs, twirls mallets round crystal bowls, and shakes rain sticks. I'd spent an entertaining afternoon round hers and the next day had ended up at a gig by Johnny Clegg, a South African musician-activist. He knew he was dying and it was to be one of his last tours. There was dancing and swaying and fist pumps and tributes to Nelson Mandela and me feeling South African and teary, even though I've only been to my parents' homeland three times. So by the time I board the train, I'm feeling a little flat. Do I really want to go to landlocked Derbyshire? Once more I have that strange feeling of contraction I sometimes feel when I head north. What is that all about? Maybe it has to more to do with being in a county that is far from the coast. Near water, I know, is where I am most at home, my truest self. Too late now, though.

On the trip up to Cromford in the Derwent Valley, I try to psyche myself up. To my great relief, Glennie turns out to be as lovely as the trees she adores: a warm, kind woman, a hugger of humans as well as trees, and dressed sensibly in a green sweater and brown cords – tree colours. I warm to her instantly. 'Welcome! Welcome!' she smiles broadly when she comes to meet me on the train platform. Were I a refugee washed up on her doorstop, I think, she'd invite me into her home unhesitatingly, brew me a tea, give me a bed to sleep in and make me feel human again.

After lunch in a rather cool warren of a bookshop called Scarthins – 'for the majority of the minorities' – Glennie takes me for a spin in her Beetle car. It is emblazoned hippie style with stickers of flowers and the words 'Wild Thing'. I chuckle when I see that, for silver-haired Glennie looks anything but wild. (Later, I recall this impression, and shake my head and give myself a telling off: 'What were you thinking? She is *all* about the wild...')

We drive through the Valley and in the rain it is a blur of sheep and quiet farm fields bordered by low, stone walls. Beyond it is the Amber Valley, and the Shining Cliff Woods. These were once medieval hunting woods, Glennie tells me and I briefly wonder if there is any connection to the 'shining light' of Elen of the Ways.

Somewhere in the woods, I know there are the remains of an ancient yew. There is a story around it: in the 18th century a charcoal burner – not a device for heating up your barbecue but a man who burned charcoal for a living – and his wife made their home inside the trunk of a yew. As infants, each of the eight children they bore were cradled in the hollowed-out branches of the tree. I only find out about the tree after I return to London, for at the time my host doesn't mention it: maybe up here the yew and the story are not such a big deal? 'In spring bluebells and anemones bloom in the woods and foxes and badgers make their home here,' says Glennie. Maybe those things are a bigger deal.

On a track nearby we park the car. We've arrived at the 'back' entrance, and I'm underwhelmed, though the black storm clouds don't help. This wood is scraggly and dense and murky. It doesn't feel majestic but confused, like a wood with mental-health issues that have not been dealt with. A wood that has lost its way. The wood at the end of my street is prettier than these woods, I think! The footpath feels like an afterthought, the sort of boundary line you might quickly walk your dog on, while hoping you didn't run into an axe murderer. It's too near to the road for my liking but at least it is lined with trees. The time has come for our walk, so I resolve to be more positive. We stop at a sycamore and Glennie pulls a face. She is not a fan of the sycamore. 'It is non-native, it suffocates all the other trees, it won't let anything grow under it, it's a "bit of a bully",' she says.

I have always felt uncomfortable with the whole 'native' and 'non-native' trees talk. If you were to replace the word 'tree' with people it would all start to sound ominous. It may feel invasive, but it's hardly the fault of the tree. It is only doing what it is hardwired to do: grow. I know that native species are a part of the local ecosystem and have lived for thousands of years in perfect harmony with their environment. A non-native tree is a foreigner tree, an outlier, an interloper. I know the mantra – I've heard it intoned often enough.

I once took part in a ring-barking at a conservation project in the wilds of Scotland. Otherwise known as 'how to kill a tree'. And who wants to do that? I had to butcher non-native trees and I hated every minute of it. I hated the blood on my hands. In the end I skulked off from the group by way of protest. I will always have a place in my heart for the refugees and the migrants of the wooded world. Besides, I think, bringing myself back to the present, I like that the sycamore is easy to identify. It has that mottled trunk and branches and its leaves remind me a little of a maple leaf.

When I was a kid I played with its sticky 'wings', peeling them apart and attaching them to the bridge of my nose (we were all at it, little nose-clipped infants). Sycamores have been around for a long time – around 100 million years. Ancient Egyptians believed sycamores connected the world of the living and the world of the dead. Some Native American nations called them 'ghosts of the forests'. Mess with a sycamore and misfortune would follow. Most touching of all is the tale I'd read in a book of folk tales. In the story, sycamore seeds are smuggled into a concentration camp in Nazi Germany. The children plant and water them with their precious water rations. With all that love focused on them in that inhumane place, one of the seeds sprouts and by the time the camp is liberated the seedling has grown into a young tree. It's transplanted to a spot where the ashes of those who've lost their lives are scattered and becomes a symbol of hope and life.

We move on to the bracken – another interloper, which 'makes the forest floor so dark, nothing else can grow.' I'm with Glennie on this one. I'm not a fan of bracken – I was surrounded by it up on the mountain in the Pyrenees and took an instant dislike to it.

I sigh. To be honest this wood just isn't doing it for me. I'm hankering after a less static, more dramatic experience, both in terms of landscape and locomotion (my own). I hide my restlessness though and next we come to a holly tree, a symbol of fertility in ancient times; in wintertime, sprigs of holly are the romantic's best friend. We peer at it respectfully. 'I'll give you some time to write down your impressions,' says Glennie.

'Even if they're fleeting?'

'Even if. Especially if,' she says. 'The real question is what do the trees tell you beyond the surface information they are giving you?'

The point of the exercise is to slow down to 'tree time'. It's about getting out of your head, feeling the uniqueness of the specimen in front of you and mulling over how you relate to it. A kind of tree psychotherapy, if you like. We might admire a tree, we might pen an ode to its beauty, but rarely do we stop long enough to pay attention to all the feelings that a tree inspires in us. Now that I am, I'm surprised at how difficult it is to do just this.

I stare at the tree, and the tree stares back impassively. It may be more bush than tree but the trunk, slender though it is, rises up ramrod straight. There is a quality of antiseptic cleanliness to its trunk and even its spiky, glossy leaves, but the tree feels unyielding, a little hard, a little distant. If this tree was a human, it might be the sort who is a bit preoccupied with appearances. I'm finding it impossible to not anthropomorphise.

The exercise reminds me of a hypnotically magical few days spent in a hobbit hut on the edge of a lake in Gloucestershire. I was there to learn how to connect with plants and weeds and flowers with an unorthodox 'intuitive' herbalist named Nathan, and I hadn't felt impatient or bored for even a minute, such was our tutor's presence. We'd chosen our specimen and reflected lover-like on its appearance, its scent, its texture. We'd given full rein to the sensations that it stirred in us. 'It's a way of soaking up the presence of the plant and becoming sensitised to its qualities,' Nathan had said. He had a way about him and managed to cast a spell over us all. He was a kind of plant magi or sorcerer. I had to tear myself away from that place, so mesmerised was I by it all.

Was Nathan's way the way of the medicine men and women of the rainforest, where plant knowledge is a vital part of life? It felt like it. Although for the healers of the rainforest the awareness is (and was) likely so much a part of their lives that it must be like drawing breath. Their knowledge is transmitted through a deep intimacy – day-to-day living, right in the heart of a wild place. Glennie, I think, has a similar bond with trees. And with the earth: 'We're conditioned to live on the surface, with our minds dominating. Well, I choose to walk a path of true and honest connection. I am the earth, of the earth and the earth cycles, the seasons are me too,' she says. 'Thirty years ago I couldn't have said that – but I get it now.'

It starts to drizzle and the gloom hangs over us. The fat raindrops smear my face and the grey makes me feel claustrophobic, but Glennie never once loses her good humour. She wants me to be comfortable, at ease. I appreciate her kindness and try harder to concentrate. We stop in front of a rowan. Its bark is smooth grey, its leaves serrated, and its berries a fire-like red. In mythology it is a tree linked with divination and prophecy. Soldiers at one time were given rowan twigs before battle to divine who might emerge the victor. 'Rowan wood has also been used magically throughout the centuries for protection and warding off evil,' says Glennie. 'If you feel the need, you can wear a sprig in your hat or make a talisman to wear round your neck.' I now see the berries everywhere, little flickering flames in the woods, lit by invisible lamplighters.

Nearby in the tangled wood is a birch, an easy spot for tree novices thanks to its silvery bark. Glennie calls it a 'nurse tree', or a pioneer tree, the first tree to move into new ground. 'Its light foliage allows other things to grow beneath it,' she says. 'Where it grows, a forest grows.' The hazel, another Florence Nightingale of the woods, is a 'generous, giving' tree, so bendy it has been used to weave roofs and boats and fences for hundreds of years. In contrast, the blackthorn, with its spiky shoots, is 'malevolent', the 'tree of ill omen' in fairy tales, but its berries are used in sloe gin. 'The only tree that I have ever felt anger in was a blackthorn. It told me to go away,' she says.

But I'm eager for something more: I ask Glennie what the secret to talking to trees is. And how do I even know which one to make friends with? 'If you're drawn to a tree, then there's a connection right there,' she says. You have to spend time with the tree, sit with your back against it, give it your attention, and the space to get comfortable with you and you it. There are no shortcuts. 'Half the time trees don't see us. We're too fast for them. Their vibration is slower.' This arcane or ancient practice (depending on your point of view) is the very antithesis of modern, manic, urban, plugged-in, neck-up, cynical, over-conditioned living. 'You have to open your heart and not have any expectations. You might visit a tree 10 times and feel nothing, and then suddenly it seems to reach out to you and you wonder if you're imagining it – but you're not. Something is happening. You're forming a relationship with it. Just as you would a person.'

You'd get closer still by studying everything to do with the tree: the stories woven around it, its history, and the ways to use its leaves and bark and fruits and flowers. A total arboreal immersion. And that is how you do it. I'm not sure how you'd know if the tree is responding – 'trust your intuition,' says Glennie – but I remember the occasions when I have hugged and clung to the plum tree in 'my' wood when heartbroken and I could almost swear the tree was cradling me. Communing with trees is not for the lazy or the sceptical.

In the rain, I feel lethargic, maybe slow and tree-like myself, so I store the information away for another time. But how lovely it would be, I think, to better get to know the beautifully scented fig in what the locals call the 'Old Orchard', in the clearing not far from the plum tree.

When we're done with the trees, Glennie tells me a little about the Grith Pioneers, a group who lived in a part of the woods back in the 1930s. They were Derbyshire's very own Thoreauvians, who craved communion with the natural world. No one knows much about them but Aldous Huxley was said to be a fan. The men were into morris dancing, wood carving and folk singing. Would these men of the woods have welcomed women? Would they have welcomed the likes of me? I don't voice the question. 'They don't live here any more,' says Glennie. 'No one lives in the woods now.'

That evening after dinner with her family in the warmth and cosiness of her cottage, she takes out her treasured tree ogham sticks. Glennie keeps the whittled pieces of twig – each from a different native tree – in a soft cloth pouch. The twigs look like wands. She doesn't say it but for some people, they're a kind of wood-hewn oracle, serious hocus-pocus, a divination tool. On the flat, smooth part of each twig, Glennie has etched mysterious rune-like markings. Each of these corresponds to a particular native tree but it's also a fiendishly complicated alphabet system created by Celtic tribes who came to Britain centuries ago. For Glennie, though, the real magic lies in the creation of the stick. 'The power is all in the intention and the making of the tree ogham,' she explains. The slow, focused work that goes into finding the right twig, the whittling, the sanding and the etching, all of it is a kind of meditation. It's the magic of artistry, with an esoteric touch. When Glennie hands me sandpaper and twigs to take home and tells me how to lovingly strip and work the bark, I know that I'm too fidgety, too clumsy to have a go. But here is a woman who is quietly keeping this ancient practice alive. That is something.

★ ★ ★

A few months later in London Glennie is giving a talk. In the room are about 50 eager faces, most of them women, most of them – but not all – young. Glennie beams: she talks a bit about the trees and about how we can heal the earth when we recognise we're part of the earth. How we can experience a connection with all life on the planet, when we practise loving kindness. 'We just need to let our veil of separation drop. It's all about choice,' she says. Her audience listens patiently. But these are urbanites who come from all corners of the world. They're climate-change activists, environmental lawyers, and some have had difficult upbringings or come from war-torn countries – there is a lot of pain in the room. In the Q&A that follows her talk, things get serious, a little heavy. 'Some people don't have choice – they have no safety,' says one woman, 'and you need to feel safe to make choices.' Another asks: 'How do I tell the trees that

I am sorry for what is happening on this earth?' Another woman says: 'My people are from Iraq, they are living with violence, with war, with horror. How can you practise loving kindness in this situation?' Only my tree-mad friend Jane, an editor at the *Guardian*, puts up her hand and asks for tips on communicating with them. Ironic, given the hostility many in the room feel towards the media.

The gentle, playful spirit I'd experienced in the woods is mostly absent. I feel for Glennie – she has left her peaceful corner of the Peak District to talk about her beloved trees and instead, everyone wants to share their pain. 'I don't have answers to all of your questions,' she says bravely.

Later, over a drink in a pub, I say to Jane: 'But isn't it interesting that all these people, with all their pain, came to a talk given by a woman who openly talks to trees?'

# A Pilgrimage Walk in a Land of Giants

Back in London, I catch up with friends and share my adventures. They nod in recognition: many have had their own esoteric experiences in some wild corner of Britain or other. It seems all I am doing is scratching the surface. There's a whole lot more going on in the land than meets the eye. Others don't understand but are curious. The lure of the magical is hard to resist. For the rest I tone it down, say I've 'been to the countryside', and we move on to common ground. Why make life difficult for yourself?

What is hard is going from my days outdoors – wandering and dallying and breathing lungfuls of fresh air in the countryside – to the kind of deadening trance that I fall under when I get anywhere near the virtual world. I worry a bit. Am I tainting myself? Am I dulling my connection with the land, with nature, with myself? I already know the answer. I know it because whenever I scroll like an addict, I feel the need to go for a swim and get clean.

I switch off my laptop. We are having yet another a hot spell, and I take to walking in the woods at the end of my street with renewed vigour. The weather is still fine and every leaf, blade of grass and branch is bursting in a sweet-sounding refrain. You'd have to have a heart of stone not to be moved by it. One clear morning, I'm so lost in thought that when I look up, I spot a new path, one that's unfamiliar to me. It's a sharp turn away from the only yew tree in these woods. How come I've never seen it before? I walk here almost every day. The path leads into the Old Orchard. It was once the abundant garden of a manor house, and now it's a kind of secluded oasis. I follow the path eagerly as even the trees look unfamiliar from here, even the leaves high above dancing in the sunlight. What is going on? I lower my gaze and come eye to eye with a small ginger fox, rooted to the spot at the end of the path. My heart skips a beat. I've never seen a fox in this wood before and certainly not one fixing me with an unblinking stare.

I long to freeze the moment, but the second I form the thought, the fox flees. I am now thoroughly disoriented. The plum tree that I am so familiar with is here but in the wrong spot. The fallen log is not in its usual place either. Then it dawns on me: the path that I've never seen before is, in fact, a path I've ignored nearly every day. I've been so lost in thought I've missed my usual turning. Yet the day I take the path less trodden (by me), I see a fox eyeballing me. There's a lesson there. And the encounter doesn't feel like a random one either. It feels like a sign, a cosmic high-five, good karma. What, I wonder, will happen if I go on a much longer walk with the same openness of spirit. What might unfold along the way? What if I embark on a pilgrimage?

A journey on foot, a slow, thoughtful, reverent walk to a special place, or to no fixed destination – undertaken alone, a pilgrimage is almost like a vision quest in motion. The inner journey is as important as the outer, you set an intention and consider what you want to honour or invoke. You could even call it a form of spellcasting, and there is something beautiful in breath and air coming together in this way. The British Pilgrimage Trust believe in the idea of a pilgrimage as something open to anyone of any faith (or none) or of any political persuasion, age or culture. Which is as it should be. Whatever way you come at it, such a walk is a chance to uncork thoughts and feelings and let them flow.

Like any walker, I know how cathartic even a 'regular' hike can be. In the Annapurna mountains in the Nepalese Himalayas, I walked for three weeks, sporting a flimsy rucksack I'd bought in a Kathmandu market stall. I had a Yorkshireman for company and suffered hellish altitude sickness when crossing a high mountain pass. I'd sobbed my way up it and down it and, emotionally speaking, I left my innards up there. I was emptied and weak, but calm by the time we reached a goat herder's hut on the other side.

Many years later I hiked for a week in Slovenia's Karst, a land with vanishing lakes – a phenomenon that owes itself to the 'porous subterranean Karst limestone'– and forests, vineyards and villages sleepier than an actual coma. The day before setting off from London I'd had a freak accident when a glass fell and shattered on my ankle. At least it had taken my mind off my navigational ineptitude. 'Beware of the bears,' said the A&E doctor who sewed up the gash. By a strange coincidence, he'd once camped in the same part of Slovenia. He told me I'd be fine. And I was. I crossed fields and woods chaotic with trees felled after a storm, traipsed softly past the ghost villages and bedded down in farmhouses. Though my sole aim had been to walk without doing myself further injury or getting hopelessly lost, I learned that even if I did, things would somehow work out. Like the time a path forked unexpectedly in three directions. I had to close my eyes, take a deep breath, feel whichever path exerted the greatest pull – and one always did a bit more than the other, even if the sensation, a gut feeling, was barely detectable – and go down it. And what do you know? It always took me in the right direction.

The most unexpected things could happen on walks: stags leaping out of bushes, antlers locked in battle; dead camel thorn trees in a desert exuding a supernaturally crackling energy; randomly sitting down for a picnic and looking up to discover you are in the midst of a tree circle, one you'd heard about for the first time just the day before; stumbling upon a medicine wheel on the edge of an Icelandic fjord. These things happened and, when they did, I felt a kind of deep, sustaining joy as if everything else in life was a dress rehearsal. But I've never done a bona fide pilgrimage walk in Britain, and so now I am rubbing my hands in anticipation.

⋆ ⋆ ⋆

My friend Olivia is flying over from New York City and she's going to join me on the St Michael's Way, a coast-to-coast walk. 'It'll be fun. We can get lost together!' I'd offered by way of enticement. She is up for it. I'd downloaded all the route links on the British Pilgrimage Trust but I need someone with me who can read a map.

And while I adore walking alone, this time I'm in the mood for company. Despite the inner satnav that has helped me out in the past, this is going to be a long walk for a single day, with not a lot of leeway for getting lost. And Olivia will be perfect company. She's a sensitive, modest, quiet, self-effacing Finn who is drawn to wild places. She writes poetry too (when she can escape the grind of Manhattan corporate life). We'll spend a few days in St Ives after our walk, which means I'll have a tiny, proper holiday too.

St Michael's Way takes one day to walk. It takes you from Lelant just outside St Ives on the north Cornwall coast, to St Michael's Mount on the south coast and is all of thirteen and a half miles. One hiker had called it a 'fancy parenthesis between seas', but it's also an abode of giants and mythical creatures. What more could I want? The walk has a certain gravitas because it's an official leg of the Camino de Santiago de Compostela, signposted with the iconic scallop shell which is the emblem of Camino pilgrims. This detail excites me no end. I've read two books by women who have walked the Camino and along the way they experienced entirely unexpected, magical happenings – strange synchronicities and encounters. The women experienced great hardship too, but given ours will be a day-long hike, this won't be an issue for us.

I love the idea of starting and ending at the sea too. It feels neat and satisfying, a tangible beginning and end. In between there'll be sand dunes and beaches, a hill to climb, views, fields and marshes to stride through, and a final, hopefully galvanising walk across a causeway to our destination.

In times past, pilgrims abandoned their ships in Lelant to avoid treacherous waters at Land's End. They carried on walking across the land to the south coast, and it is in their ancient footsteps that we will be following. All of their hopes and thoughts would be woven into the trail, like an energetic footprint or even a force field. So I see the route as a kind of liminal corridor, out of time and space, where anything can happen.

In a guidebook I read that St Michael is described as 'the Great Prince who, in the Old Testament, appeared to Daniel, defending good against evil'. Not having read either the Old or New Testament my eyes glaze over at these references. Online, I see St Michael depicted as the 'patron saint of high places'. I like the sound of that. But then I read on and what I read depresses me: 'In folklore he is portrayed as a dragon slayer, like his god-brother St George, and it is suggested that this symbolism grew out of the fact that many pre-Christian sacred sites were on high ground; the "dragons" that St Michael slayed were perhaps the last vestiges of the old religion.' The idea that St Michael was celebrated for snuffing out non-believers isn't what I want to hear. That desire to reduce the Other to dust. Same old, same old.

On the train down, Olivia and I read the Pilgrim Pledge, on the website of the Pilgrimage Trust. It's important to walk slowly; to smile at everyone you meet; to improve the way; to need less and accept more; to pass the blessing on. I can't walk any faster than a snail, so I have that one covered. I was raised in Canada with its 'have a nice day' culture and as a consequence grinning at perfect strangers is still second nature to me. As for improving the way, I'm (mostly) conscientious about collecting rubbish, and while I never find it easy to travel light, I don't find it hard to receive whatever gifts come my way. But it troubles me that I can't immediately think of a way to pass the blessing on. The blessing, I assume, is the gift of the pilgrimage walk, the land open to receive you like any warm, welcoming host.

We reach St Ives in the evening. The air is damp and the narrow streets full of puddles. Every restaurant in town is packed and we end up in a pub, eating mussels and chips on bar stools. We listen to a band with the inelegant name of 'Crunchy Frog' made up of middle-aged men who play the blues so well we idly wonder if they'd once been session guitarists for a band like the Stones. It's a fun and upbeat start, if not a particularly pilgrimage-like one. We have to tear ourselves away far too early in the evening, mindful of the long day ahead. Walking uphill to our Airbnb in the darkness, we shake the pub out of our skin and plot our intentions for our walk. After all, this is what any good pilgrim would do.

'I want the land to show me beautiful, enchanting, unexpected things,' I tell Olivia. My friend, in turn, wants out of the deadlock she is in, the life of a high-powered sustainability guru versus the poet and artist she longs to be. 'I want inspiration, I want signs,' she says. 'I don't know what to do next. Can the land guide me? Do you think it's possible?' The sky receives our intentions, absorbs them in its velvety, unknowable vastness.

Early the next morning we take a train to Lelant. Here we saunter along, cooing like a pair of wood-pigeons down a lane dripping with flowers in the brightest shades: deep-orange crocosmia, rich, blue, sweet-scented agapanthus, the dangly fuchsia – how did the universe come up with a flower all unapologetically hot purple and red? There are yellows too. A dip into my Cornwall wildflower book tells me they are coltsfoot and agrimony. This time it hurts me that I can't name the flowers, and I feel my ignorance keenly: it means I can't offer homage fully with this simple act of courtesy. I want to make amends somehow. And yet, while I can't name them, my heart is responding to the colours around me and I can feel a kind of warmth spreading inside me.

The lane leads us past some elegant homes and a cyclist riding pell-mell downhill. Eventually we reach a tiny junction, pass through a gate and walk down a path. It's lined with oak trees, like a guard of honour, and at the end is Lelant Church. This is the official starting point of St Michael's Way. The church is just one of many kinds of holy place, says the British Pilgrimage Trust. Aside from all manner of conventional places of worship,

> *'There is water in the form of holy wells, springs, river sources and confluences. There are ancient trees, sacred stones and hilltops. And there are the places where great people were born, lived, died and buried. It is for you to discover what a holy place means to you, and how best to engage with it. There are many ways to connect. No one will be watching, and no instruction given.'*

I warm to whoever wrote this, for it makes the church feel a softer, more welcoming place.

Its walls look thick, and the tombstones appear to spring out of the ferns and tall grasses. One at a time, we walk around the churchyard clockwise, to 'ritualise our encounter' as our notes suggest. Lelant is one of the most beautifully located churches I have ever come across.

Set in an enormous garden, it looks out over white sands and the Hayle Estuary beyond an incongruous golf course laced with bright red poppies. There is a tangle of tall, unmown grass behind the church. Some of the tombstones are skew-whiff, and there are oaks and sycamores and ash trees and a hawthorn leaning towards the sea like an old woman bent over double.

I sit on a bench and watch as Olivia stands on the edge of the lawn and stares at the cream sands, scribbling notes. Maybe she is composing a poem. I turn away and consider the church, and try to reflect on it as a place of 'connection with a higher truth' as the Trust puts it, a place where centuries of prayer and blessing and gratitude have created a peace that permeates the walls and transcends my unease around organised religion. I try, but I'd rather sit empty-headed in the garden and lap up the sunshine.

In this churchyard a saint called Euny was buried. He arrived from Ireland in the sixth century and, according to the British Pilgrimage Trust (who are nothing if not diplomatic), 'successfully shared Christianity with native Cornish folk.' He wasn't the first saint to arrive in Lelant; a woman called St Anta was thought to have beaten him to it. The village, which was once called Lanata, apparently took its name from her but her fate is shrouded in mystery. Was this her church? I try my best to be engaged and interested in the story and in the history of the place, but all it does is weigh me down. I find the skeleton sundial outside the church far more interesting. It reminds me of those quirky Mexican Day of the Dead figures, and the shadow it casts reminds us of time passing and the need to get going.

The first proper leg of our walk is hiking gold. It takes us back in the direction of St Ives along the South West Coast Path. There can be few more delightful walks in all the land: birds of prey hover above us, swifts carve up the sky. We follow the trail through the dunes sprouting marram grass and up along the cliff edge above Porthkidney Sands. The sands are so soothing, so empty and so creamy I gaze at them greedily, hungrily: can you devour a beach? Be satiated by it? And those flowers! Yellow gorse, orange crocosmia growing wild on the dunes' edge, cornflower-blue wildflowers I can't name, the first berries poking their heads up – we pluck and nibble on a blackberry, its purple tartness like a thimble of lifeblood. Olivia points to a knobbly tree branch that looks like it is being strangled by another and christens it 'witches' hair'. I take a snapshot and that evening I google it: it is a wisteria root.

At Carbis Bay, the sea is an intense cobalt blue with patches of turquoise, and the tourists are drawn to it in a haze of happiness. We walk past a German-speaking couple in thongs who do not return my pilgrim smile – why, I secretly wonder? Is it my face? The colour of my skin? – and we leave the path and turn inland.

Up until now the going has been slow because I am slow.

'We have a long way to go,' Olivia says, impatiently, frowning at her watch. But not only can I not go any faster, I also don't want to. How can I walk as a pilgrim with one eye on the clock?

'I won't be upset if you want to walk on ahead,' I tell her, smiling tightly, my lie obvious. After all, the whole point had been for us to do this together. Going away with a mate for anything longer than a day trip has made me feel almost normal, like a regular, non-eccentric human.

'But what if we don't finish?' says my friend. She has issues with time, she once told me. She never feels she has enough of it. Well, who does? Anyway, too bad. A pilgrimage walk is about loosening your death grip on everyday life. The journey begins and ends where we do. We define it. We breath meaning into it. A prayer made with body and land, the British Pilgrimage Trust called it. I share all of this with her. But Olivia moves as fast as a cracked whip, and she's a decade younger than me. I understand my plodding pace will hold her back. I hold my breath.

'You're right. We'll walk together,' she says, reddening slightly. I appreciate her big-heartedness, and I am relieved too – I can leave the map-reading to her. Now I can focus.

Before long, we find ourselves atop Cock Hill, facing an ugly pyramid-shaped granite shaft called Knill's Monument. Here, we cross paths with a grey-haired hiker. He clasps a hand-carved wooden staff topped with a figure of a woman. I give him my best pilgrim smile and, unlike the Germans on the coast path, he returns it. Did he carve the walking stick himself? Keen to brandish my new knowledge of such things, I ask if it's a sheela-na-gig. 'Yes, I did carve it myself but, as you can see, the woman has her legs closed *unlike* a sheela-na-gig,' he says. I am so taken aback by his reply that I nod dumbly and let him pass.

Olivia and I sit down at the back of the monument and take in the views: to the east we can see St Ives Bay with its white sands, at a distance now. Around us is heathland – a scratchy quilt of heathers in shades of amethyst, lavender and almost beetroot-purple – and patchwork fields. There is even a splash of crocosmia here and there. The little rebel flowers aside, the land here feels to me strangely devoid of spirit. Even the sky, low now with clouds gathering, seems lethargic. I take myself off to one side and listen hard: nothing. No murmurings. Someone once told me that the gods love beauty, but the monument casts a shadow over the land, so perhaps the gods have fled. I hope not, for the sake of the flowers, but I am glad to leave this place behind.

Following the rough track down, two patchwork fields on, we pass what looks like a Shire horse with giant hairy hooves, though we can't say for sure. The horse has an air of despondency about him. We croon and reach our hands across the electric fence to try to encourage him closer. It's distressing to watch him quiver and glance our way, bravely sniff us with his nostrils flared before nervously turning away and then back again. He is trying to tell us something, but we don't speak horse so we don't have a clue. Whatever it is, it isn't a happy story.

Meanwhile, out of the corner of my eye, I see four hikers enter the field. They march past, keeping their distance, two determinedly fit-looking couples with an aura of gardening and crisp rosé, certainty and National Trust outings. They nod briskly at us. Maybe my expectations have been pitched too high but, having read stories of the Camino, I thought we'd encounter more pilgrims. I imagined camaraderie, supportive smiles, encouraging nods, but there is none of this. And right now, the trail feels a little flat, not quite worthy of our attention. As if the land can hear our thoughts and is insulted, the weather changes in a flash: rain whips our faces and turns to hail. Big stones too. 'Ouch!' I yell as one bruises my cheek.

Unlike Canada geese, who waddle with their faces defiantly skywards, we have no beaks with which to deflect the missiles, and so we walk hurriedly on, hunched over. But the hail seems to rouse us all, humans and land, and like a pencil drawing turned watercolour, the flat, ferny, olive-coloured fields give way to a little thicket that – in a weird and sudden burst of sunlight – shines like an emerald, with flashes of flame-coloured wildflowers. Beyond it is a wonkily shaped stone partly covered in lichen. It is called 'Bowl Rock' and towers above us. Legend has it that it was a plaything for giants. Indeed, our splattered notes say we are entering the 'land of hilltop giants'.

Winding our way up Trencrom Hill, we briefly lose our bearings. The sun is playing hide-and-seek again and a mist has descended. It is all strange, weathered boulders, ferns and heather here. We're in some kind of no man's land – literally not for humans, but a place that feels as if it were claimed by nature spirits and fairies and giants. I have the eerie feeling we are being watched.

It's a relief to reach the top of the hill, our 'Mount Joy'. According to my Trust notes this is the 'standard medieval name for any place from which you first see your pilgrimage destination'. The view to St. Michael's Mount is spectacular. Green fields slide into the sea, and the small tidal island rises up loftily from the water. But if I turn around, I can see forest and patchwork fields tumbling into St Ives Bay too, with its white sands curving like a crescent moon. I feel a kind of ecstasy and exhilaration, while Olivia is like a feather in flight, all light and bright, the frustration of bearing with her slow friend sliding off her. We hug. We have friendship and we have *this*. 'Halfway there!' we sing.

On this hilltop, if you get here early enough and stick it out, you can see the sun rise from and set into the sea. It's like two parts of a jigsaw neatly coming together. The hill was once occupied by Neolithic farmers and later, Celts. You can still see the remains of ancient walls, and perhaps they were part of a hillfort to keep out invaders, or some kind of enclosure. But it is the strange wind-shelters that catch my eye: boulders slapped on top of each other like chubby plates. They have names like 'Giant's Hair', 'Giant's Cradle' and 'Giant's Spoon'.

I have seen boulders like this once before, in the arid desert of Hampi, a UNESCO World Heritage Site in India. I'd crossed the Tungabhadra River in an old-fashioned coracle, a glorified basket woven from reeds and saplings, and entered a dreamscape. I'd wandered aimlessly for miles, overheated and feverish in the 40-degree sun with a single bottle of water to sustain me, and the sheer number of the weird stones had given me the creeps. That night, under my mosquito netting, I'd dreamed of death and destruction, hands chopped and dripping blood. In the morning I'd shared my nightmares with another traveller and learned that in the sixteenth century Hampi had been the site of a chilling massacre. An entire empire wiped out by invaders.

This hilltop, thankfully, exudes no such horror-filled vibes. But there was once a presence here – in lore, at least: a giant named Trecobben who lived in an ancient spring hidden among these boulders. He once tried to throw a hammer to a fellow giant, Cormoran, who lived on St Michael's Mount, only being clumsy Trecobben had missed and killed Cormoran's wife. Our notes tell us that if we collect some of the spring water and pour it into the springs at our final destination, we will be engaging in a symbolic reconciliation between the giants and purging 'the violence of the past'.

'I think we should do it,' says Olivia.

'It would be a graceful thing to do,' I murmur.

'And we could do with a mission…'

'And I have a spare water bottle.'

But where is the spring? It could be anywhere.

'Inner satnav, remember?' says Olivia. Trust is key. We set off in different directions, slithering between boulders, stretching out hands and feet to see what, if anything, lies between them. 'Over here!', she shrieks after a quarter of an hour of what feels to me like fruitless searching. Perhaps it is down to the help of some hilltop sprite, or the spirit of Trecobben himself, or blind luck, but Olivia has found a recess hidden behind a giant granite rock. I join her, crouching down low and peering into what looks more like a large crack filled with water than a pool. But a pool it is.

We dip our hands into the water and then pluck some strands of wild grass, tie them carefully with a woodier reed and leave it for Trecobben or whatever undine spirit now dwells here. The grass is our version of the 'clootie' rag, the pilgrim's traditional cloth offering. Olivia pulls out an old water bottle from her rucksack and we collect the 'reconciliation' water. At the very moment we poke our heads out and turn back up the rock, a bird of prey soars overhead. A hush descends and the bird's presence feels like a blessing. I can feel the emotion pouring from it, a kind of love and wildness and wisdom. The eeriness we felt on the way up has disappeared too – as if in the act of gathering the water and beginning our re-enactment of the symbolic reconciliation, we have dispelled a stagnant shadow. Of course, we are not the only ones privy to the Pilgrimage Trust's lyrical route notes, and it may be that many a well-intentioned hiker who has come before us has been received in the same gracious way by the land. But, from here on, there's no denying that things improve.

Perhaps the land is now on our side and willing to share her secrets, for down the hill we turn into what feels like a shimmering dell, an enchanting garden of a corridor. Wandering along it, we come upon an iridescent blue dragonfly, its wings beating to a blur. It alights on a leaf and then all three of us go very still. Have you ever had a dragonfly eyeball you? It is a funny sensation, like being anointed by a tiny astronaut in a helmet, sprinkling fairy dust. Its big bulbous eyes fix on us for a long moment. 'It's a sign!' we say to each other, for dragonflies are a symbol of transformation.

Whether that belief is grounded in folklore or some other lore, it doesn't matter. In the liminal space, the place where the land metaphorically meets the sea, all things are laden with meaning. The firecracker-coloured wildflowers, the bird of prey, the horse, the dragonfly, the well water, all are part of our Songline, a living narrative. Perhaps, I think whimsically, if I walk this path again, I will sing their names and the land will rouse itself for me. In this special time and place the rational holds no sway.

The dell leads us to a grove of twisty hawthorns. 'Maybe this was once a field boundary?' whispers Olivia. We don't know what or who might be eavesdropping, though it isn't humans we are mindful of. I have no idea why the trees were there: maybe at one time they were grown and harvested for their curative tinctures and elixirs, or perhaps little people lived there. A larger-than-life storyteller up on Dartmoor, a man who was robust and theatrical, once stood over a hawthorn and said in his rich, mellifluous voice that the tree was a gateway to the fairy underworld. I'd read, too, in a book of folk tales that if you pluck a thorn from the tree, then you might piss off the fairies who live in the roots. If this happened, you'd be pulled into their world for an unreasonable length of time. It would feel like the blink of an eye, but if you found your way back to this world, you'd discover to your horror that many decades had passed and your loved ones were long dead. I whisper the story to Olivia, and we walk on quickly, just to be on the safe side.

At the top of a rise through a dark, gloomy forest – I christen it 'goblin territory'– it is a relief to emerge into the grounds of Ludgvan church. Inside on a table we find two bottles of orange squash and paper cups, a kindly offering for any passing pilgrims. I'd rather water, but we're thirsty and help ourselves. I scribble a thank-you note from 'two thirsty and grateful pilgrims' and leave it on the table.

A vicar had once prayed for a miracle in this church, and in response, the Divine offered a spring, flowing from the ground. The water, the vicar discovered, gave you sight and made you eloquent. It has long disappeared, though that doesn't stop us from mooching around and trying to find it. 'Will you be the pilgrim who makes the water flow today?' our notes ask us teasingly.

By now we've been walking for nearly nine hours. I am desperate for a hot drink but I practically have to drag Olivia to the pub next door for a coffee. My friend is unstoppable, a kind of human arrow to my feather. The White Hart is cosy, and I am grateful for its warmth. It is one of the oldest pubs in Britain, all low beams and winding passages. In a nook there is a ledge full of books, which makes me feel this is a friendly place: important, because in rural pubs mine is often the only non-white face and I always cross the threshold of one with slight trepidation (that feeling is my normal, it's never *just* about what's on the menu or whether it looks cosy). The symbolism of the White Hart isn't lost on me either: in Arthurian times, the stag was a sign that it was time to embark on a quest, but in lore it is also a symbol of purity, a healer of the land, a protector, the king of the forest, a sacred creature never to be harmed.

The final leg of our walk is through the Marazion Marsh, a nature reserve. Crossing a busy highway to get to it, we catch glimpses of St Michael's Mount, now looming. I am slightly awed to see it up close. The clouds begin to blacken as we cross the boardwalk through the reed bed. I hold my breath in the illogical hope of keeping the rain at bay but the heavens open and the wet stuff comes down in sheets. There are no birds, no waterfowl in sight. Just as well neither of us is a box-ticking birdwatcher. To my relief, I feel a renewed burst of energy and my legs are doing well. Olivia, naturally, has shown no signs of fatigue all day. As we leave the marshes, we can see the dunes and – at last! – the causeway to the tidal island. But now the rain has turned to hail again. No-holds-barred, golf-ball-sized ones that pummel us – nothing like the soft, rebuking slap on the cheek of earlier. The storm is reaching a crescendo, the elements finally free to unleash their full power, having courteously held off till we've made it.

A sign on the beach says that the causeway will be open from 7.10pm. I check my watch: it's 7.15pm. 'How's that for synchronicity?' I say to my friend, a little giddily. Despite my sloth-like pace, we have somehow managed to arrive at precisely the right moment. Any earlier and we'd have had to sit and shiver and wait for the tide to recede. Now, though it's safe to cross, the waves are still billowing on either side of the causeway. A father and son nod as they pass us in the opposite direction: the boy is barefoot, the elder's shorts soaked. The parting of the sea has been an illusion, and the water sloshes and slithers over our boots. We are not walking barefoot as a true pilgrim might do, and the soaking feels like a telling off. Then, the minute we set foot on the island, the squall stops dead. It is uncanny. 'Like the island knows we've arrived,' remarks Olivia.

Mount St Michael seems more like a manicured tourist attraction than a pilgrimage site although there are few people about given the hour and the weather. We immediately set off to find the giant's well to pour into it our precious Trencrom Hill water but, to our dismay, the path leading to it is behind a locked gate. Somewhere up ahead is also a heart-shaped stone embedded in the ground, which the legend says is the heart of Cormoran, broken when Trecobben accidentally killed his wife. All of it – the castle, the chapel and the gardens – is off limits to us. As too is the spot where Michael, the archangel who guided fishermen to safety, kept watch. But we have come in good faith and so Olivia pours the water over the fence anyway. We have made an offering of peace, we have enacted the symbolic reconciliation of the giants. We don't know it yet, but this simple, childlike act of ours will very soon be acknowledged in the most unexpected of ways.

⋆ ⋆ ⋆

It happens the next evening. It's been a wet day, and we've visited the Barbara Hepworth sculpture garden and the Tate Modern. We are killing time before dinner and find ourselves on St Ives island. Until we stumble upon the path to it, we'd not even known it was there. It isn't really an island at all, but a peninsula. Migratory birds are fond of it as are dolphins and porpoises who like to frolic in the surf – although at the time we don't know any of this. There is a small chapel here too, the chapel of St Nicholas, the patron saint of children and sailors. There is another, female saint associated with the island, St Ia, the founder of St Ives. She was an Irish princess, and brother to Uny, of St Euny's church in Lelant. In the British Pilgrimage Trust's version of the tale, she was eager to go to Cornwall and tell the locals about God but was too young to sail. One day she spied a green leaf on the sea, magically made it grow and drifted across the sparkling waves on it till she reached land.

The day has been a wash-out, but now finally the rain lets up. The low sun bursts out from behind the clouds, the surf crashes against the black rocks, and the wind is pushing against us in an excitable way. The promenade is a sensational spot, wild-feeling yet only a short walk from the town centre. We begin walking towards the chapel on the hill, and halfway up we pause to enjoy the bay views. There is nothing between us and Canada but the Atlantic. By boat it would take weeks to get there, weeks of nothing but waves and seabirds and loneliness and private thoughts and no thoughts and listening to whatever language the sea speaks, and communing with the stars in the sky. You'd have flying fish and dolphins and porpoises for company and beneath you strange and unknown sea creatures. You'd experience wonderment at the vastness of the universe and humility at your infinitesimally tiny place in it. It's a tempting thought, an old-fashioned sea journey, a pilgrimage of another kind for another time.

There's a body floating in the water. We freeze. On closer inspection, it's not a body, but a dark, smooth *something* bobbing above the surface. We peer intently into the surf, trying to get a handle on it. 'Did it just move?' says Olivia. As if in response, the thing rises up and throws itself back into the water with abandon. 'Oh, it's a dolphin,' we exclaim excitedly – the marine mammal seems more splashily playful dolphin than a shy porpoise. It surfaces once more before sinking under the water. We will it to reappear and then it does, leaping clear of the waves. What was it I'd read about the intelligence of cetaceans? Something about the size of a dolphin's brain being bigger than ours?

While I am marvelling at it, laughing with it till it disappears – dolphins have that effect on you – another small miracle is unfurling. When I look over my shoulder, Olivia is no longer by my side. She is at the top of the path, waving furiously at me. What has she seen? I jog up to join her and face the other side of the island, and I see it too: a double rainbow, clearly visible from arc to arc. It starts in the sea and seems to end precisely at the church in Lelant. I've never seen a double rainbow before. This one is so bright, its colours so vivid and shimmery, that it appears less a weather phenomenon and more some kind of celestial art. It spreads itself across the sky, flirting with the blue ether. What is really odd is no one else is taking the slightest notice of it, even though you can't exactly miss a double rainbow. A Japanese couple smiles at us because we are squealing like small children, but they don't so much as glance at the sky. Ditto the two teens, hunched over on a bench and sharing a bottle of Coke. Far down on the beach below, it is much the same: a man is paddle-boarding in the water, people are lying in the sand, a toddler is building a sandcastle. No one is pointing or staring at the sky. Not a single person. If I were alone, I'd think I was imagining the rainbow. But I'm not alone.

'Why isn't anyone else seeing this?' says Olivia. How often do double rainbows appear around here, we wonder? Are they a daily occurrence? So familiar as to be unremarkable?

'What if we're the only ones who can see it?' I say, sharing aloud the lunatic thought.

My friend nods thoughtfully: 'Well, we did make the pilgrimage,' she says slowly, like me, caught up in the magic. 'And we did carry the spring water to St Michael's Mount,' I add.

The walk has sharpened our attention, primed our perception. We are still in a thin place, where the walls between human, animal, element, matter and spirit are gossamer, and where highly charged encounters between species are perfectly normal. In such a space and place a double rainbow as an offering of nature – a bouquet emblazoned across the sky for us – doesn't seem entirely outlandish or ridiculously egocentric. For isn't nature capable of miracles? There is a dimension to our experience that feels indelible and out of the ordinary. Had we not paused to watch the dolphin at that moment, we'd have missed the rainbow: divine timing is everything. The British Pilgrimage Trust had said: 'Don't hide your experiences under a bushel.' No chance of that.

# Lost in Glastonbury

One minute I'm eyeing up soft, buttery Somerset fields and hills from a bus window, the next I'm deposited at the foot of the town's High Street and I have to resist the urge to flee. It is intense in the extreme, like finding yourself enveloped in a cloud of too-strong perfume, although here it's industrial-strength incense.

The shelves and racks and window displays in the shops are lined with books on miracles, the afterlife and mythology. There are crystals and wands, statues of goddesses and gods and pagan figures, packs of tarot cards and runes, capes and chalices, shimmery mystical paintings, good-luck charms, healing potions and CDs of floaty harp music. It's not that I'm averse to any of this – except maybe the harp music – but I feel like I've eaten one too many sweets. There's too much conviction, no room for doubts. Otherness takes centre stage here and I'm overwhelmed by it. Or maybe it's the rampant consumerism.

Glastonbury has been a pilgrimage place for thousands of years, steeped in mystical lore. People from around the world make a beeline here – in search of acceptance and solace and a certain kind of safety – because of the town's inclusivity and celebration of all things alternative. Others come because of its connection with the mythical Avalon. They are eager to make the pilgrimage to its revered sites: Glastonbury Tor, Chalice Well, Glastonbury Abbey. But what does any of this have to do with me?

It's a question I ask myself as I walk dizzily past the shops and the signs for healing and mystical transmissions and goddess workshops. Why have I come here? Why am I not in the woods, in a forest, in the wild, on the coast? Why am I walking down a street, looking at gargoyles and halogen lamps and talismans? The only patch of grass I can see is in the courtyard of the church, occupied by men and women with pinched, prematurely lined faces, clutching cans of beer. I'm here, I remind myself, because I know there will be other humans here who believe, like me, that the land has a voice, a spirit that seeks – or is at least open to – a connection with us humans. I'm craving solidarity, I suppose. And understanding. I was born on this land, and while I walk on this land, I want to connect with the spirit of this land. In my own way. Here, in this place of all-comers-welcome, I'd like to think there will be others who believe that such a thing is possible. In Glastonbury, everything is open and in the raw, no editing your beliefs to make them palatable. No fear of sceptics because no one here engages with sceptics. The only Other likely to rouse anyone and get their blood pumping is the non-human kind. At least, that is what I like to think. Maybe I am being naive. I'm a tourist, I'm an outsider, I'm flitting through and I can't possibly gauge the undercurrents here. Yet in Glastonbury, with its mishmash of cultural myths and beliefs, I figure I ought to find a natural home. I may be among friends but I am in at the deep end.

In an airy-looking vegan cafe I sip on a 'Chalice Chocolate' – it's made of cashew nut milk and raw cacao – and I form a plan. I want to wander up to the Tor, to experience whatever it is that draws the locals up here over and over again. It's a start, right? After that, I'm not too sure. I'll follow my nose; it hasn't let me down so far. I hope that this place will come alive for me.

In the evening, I meet my friend Will who lives just outside the town. He has the air of a slightly distracted, bespectacled professor – only he's not, he works in public relations. Will loves his festivals, his singing and his workshops, but he can step outside of the bubble. 'Do you fancy a stroll?' he says. A stroll would be good. I'm desperate to stretch my legs.

After dinner (in an Indian restaurant where the Indian waiter glares at me – why, I don't know, seeing as I'm the only other person with a brown face in here) we walk past my Airbnb and Will points out a shortcut to the Tor. It's past an ashram with Tibetan prayer wheels, glittering beneath the starry night sky. We're only a few hours out of London and I'm astonished to see the stars. What we miss, in the city, with our veil of pollution. We spin the wheels clockwise to send prayers out to the world as per the custom, then we walk back slowly and I tell Will I'm not too sure what I'll do after I go up the Tor. 'I have an idea,' he says. 'I have a friend. She is a priestess of the Goddess. Would you like to meet her? Maybe you can go for a walk with her around Chalice Well Gardens tomorrow.'

I'm not sure what a priestess of the Goddess is or does, but Will tells me Goddess culture is bound up in the cycles of nature. 'I'll let Heloise tell you more,' he nods before walking me to my door. It all sounds promising, I think as I crawl into bed. A plan.

* * *

The next morning, I wake gritty-eyed. I'd slept badly on account of the Christmas tree lights strangling the giant conifer outside in the garden. They'd blinked on and off, hour after hour, like red devil eyes, which I'm sure someone in this town would have something to say about. Over breakfast, I chat to my hostess, who owns a pub. She has dabbled in Goddess culture but turned away from it because there is 'too much politicking,' she says. She sits down and shares with me her conspiracy theory called 'Agenda 21'. This conspiracy is something to do with sustainability leading to global dictatorship, a kind of eco-totalitarianism. 'They're all in on it: Obama too,' she says, all intense, middle-class zeal. Is this why she keeps the Christmas tree lights on all night long? As a kind of protest against all the sinister Greens? I smile politely and eat palaeo muesli and sip herbal tea but I'm confused – this conversation is not what I was expecting first thing in the morning and not in Glastonbury of all places.

Fortunately my mood lifts when I step outside, released from the odd conversation. It is a brilliantly clear day, and the sun is bright and perfect, not searing but warm enough to soothe. Expectation weighs heavily on my shoulders as I walk past the ashram. I'm too self-conscious to spin the prayer wheels in broad daylight, though I peer into the ashram garden. It looks peaceful and I am guessing there is no conspiracy talk here first thing in the morning, although who can say what goes on behind closed doors? I squeeze past a narrow entrance, and walk up and along the footpath into the fields. A gate takes me past a field where a curly-horned ram and a few sheep are grazing, and then into another field beyond it. As I walk, my thoughts turn to the Tor, which I can't yet see. The 13th-century tower – all that remains of a stone church – rises above the flat fields of the Somerset Levels like a weathered celebrity. Thanks in part to the terracing around the slopes ('seven spiralling paths that some say may have once been a Neolithic labyrinth that guided pilgrims up the conical hill,' according to a guidebook I'd read in the B&B), the Tor exudes a kind of mystical glamour.

No matter how much you care to look upon this as a pleasant country walk to a pretty viewpoint, thanks to the Tor and the stories that swirl around it, it will never be that alone. The myths inspire an expectation of Otherness that exerts a great, magnetic pull – it's a cosmic attraction. Hermits and monks, pagans and Christians, even fairies have made the Tor and its surroundings home. I'd heard that hidden caves lie beneath it. Some say there was once a Neolithic stone circle at the top of the Tor. And that it was surrounded by a lake, and known as the mystical isle of Avalon.

I come to another gate and a long, tree-shaded lane and more fields. They begin to roll one into the other, like the myths. I can see the Tor now. There is something so restful about the lay of the land and the spiralling path and tower and the sunshine that I sigh in delight. I am pleased with myself, pleased with the effort I've made to get here.

When I'd said I was coming, a healer friend in London who had a fondness for Glastonbury told me a story about the mythical Avalon and the golden apples that grew there. 'If you were offered an apple in Avalon your powers of perception would heighten, and you'd gain the gift of sight. Your life would be for ever changed, and you'd be connected to the Divine – or some kind of cosmic intelligence,' she said. Some people believe that the Holy Grail – the cup of Christian and Celtic legend, imbued with miraculous, healing powers – was hidden in the Avalon orchard. This too was part of the tale, for if you picked a golden apple and put it in the Holy Grail, it'd turn into a magical elixir. Sip it and you'd be immortal and whole: loneliness begone! The isle was a kind of utopia symbolising bravery and abundance and magic.

I love this slice of Somerset, I think to myself fiercely. If I'm not a local, am I allowed to love a place as my own? As I walk through the fields, I daydream: if I move here, a more peaceful version of myself will take root. Maybe I'll spot a large blue, a rare butterfly that can only be found here, a starburst of colour with its black spots and mauve-purple hue, meant to be a harbinger of good fortune. On a visit to India's Meghalaya state a few years before I'd disturbed a whole vale of butterflies at the foot of a deep, forested valley. The giant yellow ones had danced round my head in a halo and the scent of orchids had left me woozy. Mushrooms jutting out of trees like hands, black and white spiders as large as dinner plates, long green caterpillars, I'd felt like Alice in Wonderland with each new discovery. Could such an exotic fever dream exist in England, even in a place as dreamy as Glastonbury?

At the base of the Tor, I see the sign for the 'Avalon Orchard'. I'm tempted to follow the arrow but when it comes to it, I lack the courage to pluck a piece of fruit from a tree – it feels impolite, somehow, more than mere scrumping. Do I have the right to pluck an apple from the Avalon orchard in the way that a local might? Or maybe a local would never do such a thing. A friend in Shropshire has an orchard and I remember him telling me that he only grew cider apples. 'You wouldn't want to eat them,' he'd said. So maybe the apples in the Avalon Orchard are cider apples too, and taste nothing like a bite of a succulent, juicy Avalonian golden apple of myth.

Anyway, I'm here to climb the Tor. I can see a tiny figure at the top, so that means I won't have it to myself. But at least there is no one crowding me on the walk up. 'Baa!' I say to the sheep who stare boldly but then move away when I get too close. I stop halfway up and drink in the views: quilted fields and rooftops.

When I reach the top, I walk round the tower. It looks a little forlorn, if such a thing could be said of a tower. Like a matriarch who has lost every member of her family but endures, straight-backed. The National Trust says the Tor is the remains of a 14th-century church. If I want to, I can walk through it to the other side, but inside I can make out a hooded figure sitting hunched over on a bench. He looks like a monk caught in the mists of time, and though I'm curious, I leave him to himself. Instead, I peer at the fields below and breathe in the fresh, clear air. Now what?

The locals, I imagine, come up here when the dawn light seeps over the land or when dusk sweeps the sky. Maybe a few even sleep out here, hold ceremonies here.

They have forged a relationship with the Tor through repeat visits, their bond unspoken but heartfelt. Maybe they see or intuit things, but all I feel is the sun on my face and pleasure in the pastoral beauty around me, so sweet I feel it enter my heart, rise up in my chest and want to escape through my mouth in a kiss. I look back to the tower and see the man in the hoodie walking away – he's not a monk stuck in time, after all – and so I go in. The first thing I notice is a big illustrated book on dragons on the bench. I'm not sure if it's a children's book or a book for adults who are into fantasy fiction. A gust of wind blows in and the page opens randomly onto a riddle:

*It is cold and it is hot*
*It is white and it is dark*
*It is stone and it is wax*
*Its true nature is of flesh*
*And its colour is red.*

*The answer is: the human heart.*

I peer at the riddle as if it's going to provide the answer to a question I haven't yet asked. I have been living by clues and synchronicity and wisps of threads of trails, and now they have become my bread and water – though, given that I am mesmerised by a book on dragons, perhaps I need to vary my diet a little.

Still, I read the words slowly. I read the words fast. And nothing. What about the dragons? They symbolise nobility, good fortune, strength, courage and power. The dragon is my sign in Chinese astrology. 'You have an unquenchable fire,' a soothsayer said to me once. (I took it as a compliment, but I think he may have been making a tactful reference to my temper.) I read somewhere that the ancient Chinese considered themselves to be the direct descendants of dragons. A barrister-turned-earth-warrior I knew believed unashamedly in dragons. She'd shared a fantastic tale about how dragons were stirring under the land, ready to unleash their power. In some parts of Eastern Europe, she'd said, there is a belief that dragons are protectors of land and people.

Not a lot else happens up on the Tor, but the walk has done me good and blown away the cobwebs, and as I walk back into town I have high hopes for the afternoon.

⋆ ⋆ ⋆

I check my watch. The mystery priestess is running late. I go over what Will told me when I'd rung him at lunchtime. 'Her name is Heloise Pilkington and she's a singer and a bit more than that too.' She was once a folk singer, and then part of an a cappella trio but the 'bit more' relates to a more esoteric dimension to her singing. She uses the vibrations of her voice and objects like tuning forks, gongs and Tibetan singing bowls to stimulate healing – a bit like my environmentalist friend who did the overtoning. Will said she grew up privileged, in London next door to a Beatle.

'Which one?' I'd asked. I couldn't help it.

'Paul McCartney, I think.'

Whoever this priestess of the Goddess is, I hope she will lead me by the hand, metaphorically speaking, into the beating heart of Glastonbury. I'm impatient now.

She arrives in a graceful swoop. 'I'm sorry, got stuck in traffic,' she says, shaking my hand. Heloise is a tall, elegant, self-possessed redhead with pixiesh features. 'The land here is extremely powerful,' she says as we enter the gardens of Chalice Well through a pergola dripping with fragrant wisteria, climbing vines, and roses.

The gardens are the site of a holy well. It's a peace garden and a healing sanctuary. That much I know. This is not my first visit; I came here a decade ago on a miserable, wet day to meet a woman named Natalie who was an energy healer. We walked up to the Tor and made strange symbols with our fingers and arms and legs. 'Mudras', she called them. I was embarrassed not to know the word, being Indian, because it's a Sanskrit one. Anyway, I was soaked to the skin and kept my head down. I had no sense of where I was, either geographically or culturally. I don't think I even knew which county I was in. I was so open to the world, and yet so naive about Britain.

While I was waiting to meet Heloise I'd picked up a leaflet at the gate. It told me the waters in Chalice Well are 'the essence of life… a direct expression of an unbounded life force.' I have tried to get to grips with the legends surrounding the famed iron-oxide-rich Red Springs that flow here, but every time brain fog descends. Joseph of Arimathea, a disciple of Jesus, had the Holy Grail in his possession, and caught His blood at the Last Supper. Joseph travelled to Britain with the Grail and buried it below the Tor, where the Red Springs began to flow. I feel weird telling the story because it's not mine and wasn't handed down to me. And I don't know for sure, but I think maybe the Christian myth is one a priestess of the Goddess might take issue with.

Heloise smiles politely when I share all of this. 'Hm, yes, the *Christian* myth,' she says. She tells me another story, one in which the Red Springs represents a feminine energy. There's an association with menstrual blood and a woman's creative power. 'Priestesses connect with the feminine aspect of the divine, the Goddess. Some might say she is the entire living earth,' says Heloise as we amble among the yews, through the meadows, the pools and flower-lined paths. The earth as a mother who has given birth to everything we need to survive on this planet – I like that.

Being a follower of the Goddess is also about having a more lyrical relationship with life, full-stop, injecting poetry and beauty into it. 'But what does a priestess do exactly?' I ask. The vague ideas I have centre around ethereal robes and Kate Bush-style dancing that I am unable to edit out of my mind's eye.

'We create ceremonies and gather with like-minded women to express our devotion to the various nature deities by chanting and singing and making prayers and offerings,' says Heloise.

Illogically, I feel a little let down that singing and chanting are involved. I am a bit iffy on chanting and I can't sing. But the rest of it: cool. Later, reflecting on all of this, the words of an ecologist who'd I'd met briefly in Devon a few years back, will come to mind. 'It's OK to feel what you feel in the forest… what you feel or intuit out in nature or by the sea is telling you something… all of that is a genuine form of communication from the earth,' he'd said. He was trying to breathe an animistic spirit and language into the language of science. He knew how to bring the earth alive for those who struggled to do so. It was the same for a priestess of the Goddess. 'Gaia's a dirty word in science, but it's changing,' the man had also said. He may have been up against it, but he was an ecologist, so people listened to him. If you were a priestess maybe they would, maybe they wouldn't.

★ ★ ★

I can understand why people gravitate to the gardens: not only are they beautiful, they're restful and inviting too. Maybe they are animated by a benevolent mothering spirit. Whatever, it's a seductive embrace and it pulls me in. We join some meditators by the well head with its wooden lid inlaid with wrought iron. The symbol on the lid is of two overlapping circles, a vesica piscis crossed by a rod or a staff. Around the circles are curling leaves and flowers.

The Latin words are new to me, though the two black interlocking circles look familiar enough. They symbolise the sacred union between opposites – the inner and the outer worlds, heaven and earth, spirit and matter, yin and yang. This vesica piscis was designed in the early 19th century by one Frederick Bligh Bond, an architect and archaeologist with mystical leanings. He was in charge of the excavation of the ruins of Glastonbury Abbey – and reportedly had some help in his work from long-dead monks.

After sitting quietly for a while we walk down to the Lion's Head drinking fountain. The water tastes sparkling fresh, alive, and faintly like the wild, iron tang of blood. The mystics say it has miraculous healing powers. I'm not sure what I believe, but I fill up my water bottle with it to be on the safe side. Next, we walk over to the two yews near the exit, and place our hands on the ancient trees. These trees are believed to induce visions, to help people take journeys into other realms. It's the tree of immortality and the death tree – the one whose berry seeds are toxic. I'm disappointed that I feel nothing, but the stories work their magic. 'The Goddess is present everywhere, in these trees, in the flowers, the water, the orchard, the meadow, the sun above us, all of it,' says Heloise.

Various aspects of the Goddess are celebrated on special days through the year, Heloise explains. I understand this from my mother's Hindu faith – Shiva, Vishnu, Kali, Lakshmi, they are all aspects of the Divine. I even have a shrine to Lakshmi in my bedroom, put there after my mum had told me that if I wanted to make any money I'd better start paying my respects to the goddess of wealth and abundance. I didn't need much encouragement.

We lazily backtrack past the various pools. There is one that is a bit like a shallow spa pool, another with 'bowls' that recreate the swirling eddies of a mountain stream. The sounds of the tumbling water all around us are soothing. Through a willow arch we enter a meadow with views of the Tor. Young lovers sprawl on the grass, their limbs entwined, while families picnic here. Nearby is a waterfall hidden under the shade of another yew, the sunlight catching the water and sparkling. I see a stone seat over which an angel keeps watch. There is a physic garden here too with arnica, valerian, mugwort and other plants.

I'm grateful to Heloise for the tour but I can't wait to return alone to the garden the next day. Maybe I'll come back every year. If nowhere else, at least in this garden in Somerset there is a place that welcomes me exactly as I am, and that I, the outlier, can call a home. Maybe this is what I'm looking for: little pockets of home, all over Britain.

'Would you like to see the White Springs Temple?' asks Heloise. 'It's only five minutes away.' I leap at the opportunity for I've heard much about this temple, named for the calcium deposits in the water that leave a whitish trail. It's in a reservoir just up the road, and it's the yang to Chalice Wells garden's yin. Sacred place it may be, filled with pools and shrines and altars to various deities, but it has an altogether darker energy. While Chalice Wells is lightness and sunshine and tranquillity, the White Springs Temple feels like a spooky cave.

I hesitate to cross the threshold: it feels too raw, too murky, like a portal to an underworld. Maybe it's a place to make peace with your shadows? But I don't fit in here. I'm not into naked bathing, I don't have piercings in my nose, I don't make medicine from herbs, I don't smoke dope, and I hate drumming. I also don't dance crazily under a full moon and I prefer to keep my clothes on in public. I'm not ostentatious about my eccentricities, and I express my 'wild' inwardly. Yet, at the entrance, part of me feels like a moth to a flame.

A thin, nest-haired temple guardian smiles broadly when we enter. I can flee now, I think. But part of me doesn't want to flee. The flickering candles draw me in and Heloise sweeps me along. The temple is like a carousel of the night, a place reflecting the murky depths of the human soul, the suppressed chaos that lurks within us all, the deviancy, the fear. What goes on in here, I wonder? I'm not sure I want to know. 'Come,' says Heloise. 'I want to show you something.'

She leads me to a gloomy corner. Here, I make out the outline of a brilliant orange flame, only it morphs into a painting, jagged concentric circles in shades of orange and red, above which is a pair of hands clasped around a flame, and a faint-looking face. On the altar are a candelabra with three candles, and white roses in a white vase. Heloise tells me this is the goddess Brigid. 'She's the keeper of the flame, the sacred guardian of springs and wells. She represents wisdom, inspiration, healing. Like Saraswati,' she whispers.

Well, sort of. The Hindu goddess is a symbol of knowledge and learning and is beloved of parents willing their children on to greater and greater academic heights, parents who sometimes conveniently forget that Saraswati is also a goddess of the arts, of music, of wisdom. 'Saraswati was originally a river goddess too – her name means flowing and watery,' I tell Heloise. Saraswati is also associated with the lotus, a flower I have tattooed on my back for reasons that have nothing to do with the goddess, though I like the association. On auspicious days some devotees would drape a statue of the goddess in silks, and would bring their books and musical instruments to her to be blessed. My own mother made sure the first book I wrote – the only other book I have written – was blessed by Saraswati.

(Next time I think I'll make sure Lakshmi does the honours too.)

Brigid feels to me a more forgiving deity, less demanding of perfection. Or maybe that's because when I was younger I was told by my mother to pray to Saraswati if I wanted to do well in my exams. I can still remember on very rare occasions being dragged round a Hindu temple, not knowing what anything in it meant. I'd ask of every garlanded deity I'd pass: 'And what does this *mean*? Explain it to me, I don't get it.' But my mother couldn't because she didn't really understand the specifics either. Her vagueness about such things frustrated me, though I can't help but think I've inherited the quality.

The Saraswati analogy is apt in other ways, for the White Springs Temple reminds me of those sumptuous, exquisite, yet simultaneously murky, sticky-floored temples in southern India where you tread carefully, avoid the kohl-lined eyes of vaguely sinister-looking, half-naked priests, and inhale clouds of overpowering incense, as murmurings and bells and chanting envelope you.

Under the high-domed, vaulted ceilings here, we sashay along slowly, almost processionally, because the temple is crawling with visitors. A bathing pool comes into view and as my eyes adjust to the light I can see a slender, naked woman walking slowly around the edge, like a queen or an empress. Then she plunges in. I'm not sure if I admire her or worry for her. Near the pool sits a queue of men in various stages of undress. I'm no prude but right now I feel uncomfortably like a voyeur.

As we tiptoe across to other side, we enter another shrine, right by the back. There's a large painting in front of us that depicts a man with a deer mask over the top half of his face. In one arm is a hawk or a falcon. In his other arm he is carrying a hare. 'It's the Horned God,' says Heloise. 'He is wildness, raw sexuality, the hunter, the embodiment of the sacred masculine – the masculine expression of divinity, the Lord of the Underworld.'

The image is striking and the painting fizzes with energy. I feel the eyes of the Horned God on me. It's a two-way thing and I stare back and feel myself pulled in. Is everything in this temple watching me? Does creating an image of a deity potentise it, bring it to life? Heloise must feel something too, for without a trace of self-consciousness she breaks into song. Her voice is high and clear and crystalline. Is her song an offering for or a celebration of the figure in front of us? Maybe it is both of these things.

Later, after a glass of wine, I'll write about the temple in my diary in a shamelessly melodramatic way: 'It feels like the thorns that scrape your skin, but let you know you're alive, the roots that trip you up and pull you out of your trance, the sea that drags you into its depths, terrified before you surface, like a newborn.' It's only been a day and I already sound like a local.

'Well, that was something else,' I tell Heloise when we're back outside, blinking in the sunlight. I follow her over a fence and up to Chalice Hill. On the gentle slope, we pass a grove of oak trees, the frilly leaves like pieces of a jigsaw puzzle.

'Only a little bit further,' she says mysteriously. I feel I am about to be initiated into something Goddessy and special. We keep walking till we reach a circle of tall, sturdy beech trees. 'So tell me about your love life,' she says, to my surprise down-to-earth and girlfriend-like now.

Apparently this is a good spot for exchanging confidences: the beech tree is a queen of the woods, a protective, nurturing tree, a wise tree. So, here in the circle of trees, I spill and we talk, woman to woman, about the men who have come in and out of our lives. She is in a relationship with an older man who she loves but needs to break away from. I am single and I am not happy about it. I've been alone too long.

I wonder if ours is a conversation women always have after a visit to the Horned God. But in this landscape, Heloise tells me, a figure of a reclining lady can be made out if you have the eyes to see her. 'The beech trees are the belly, linked to fertility and the feminine,' she says. Maybe that is why this is the perfect place to talk about a boyfriend who cannot let go and a woman who cannot find someone to hold on to.

The thought suddenly strikes me that my desire for intimacy with the land might be a displacement activity for a lack of romantic love and sex in my life. I've been lover-less and boyfriend-less for longer than I'd care to admit. I carefully squash the thought down, because if this is my motivation it diminishes everything I'm doing. Anyway, I don't believe it is. It may be part of it – I'll concede that – but not the whole. And even if it is, I'm sure I'm not the only human who has sought solace in the land because they want to feel whole again, whatever it is that has left them broken.

I remember back on top of that mountain in the Pyrenees, I'd been overcome by a strange urge to couple with the earth and had pleasured myself – whisper it – naked under a cloudless sky. Afterwards I hadn't bothered to put my clothes back on. What was the point? There was no one around for miles and it was baking-hot that August. I think a priestess of the Goddess would have approved.

In the evening I listen to Heloise sing in a small ceremonial room, through an alley off the High Street and up a flight of stairs. Propped against the wall, surrounded by cushions and curious seekers, I struggle to stay awake. In the end I let the sounds flow over me – she certainly has a supple voice, like the Goddessy version of free-flow jazz, and I doze, drowsy after the long day.

⋆ ⋆ ⋆

The next morning after hanging out in the gardens, I'm waylaid by a small art gallery. There is something about the smallness of it that intrigues me. It is intimate and welcoming and filled with paintings of stags and slender goddesses draped over tree branches, angelic figures with wings, masked figures and mystical creatures, detailed woodland scenes that look as though the painter has conjured them from his dreams or perhaps a long spell in a deep, green forest. The gallery is owned by a man called Yuri Leitch, who is also the artist. A biographical note tells me he is interested in Celtic and Arthurian traditions. He is a writer too and the editor of *The Signs and Secrets of the Glastonbury Zodiac*. What, I wonder, is the Glastonbury zodiac?

I've seen a copy of this book in my bedroom at the B&B but haven't had time to flick through it. In the gallery, I hang around pretending to look at the paintings just so that I can chat to him. The artist, who is sitting in a corner reading a book, is a bear of a man, tall and bearded. He looks like a sage or a magi, the kind of man who maybe won't suffer fools gladly. I circle around awkwardly for a bit, and he pretends not to notice. I'm not entirely sure what to say or even how to say it but I want to share something of my hunger to connect with the land in a more intimate way. I have a feeling he'll understand.

'I want the land to speak to me, to guide me. It *is* guiding me. I know this because when I enter into this spirit, interesting things happen. But how do I know what is true and real and what is my imagination?' I say, blurting out the words and feeling foolish. What must this man be thinking? But then this is Glastonbury. If I can't voice such thoughts here and to this man, where can I voice them?

Yuri Leitch considers the torrent of words he's meant to field. He chooses his own with care. 'You need to let your empathy and integrity determine whether you are genuinely experiencing a dialogue with the land or merely projecting your own fantasies,' he says. There is kindness in his voice.

*Empathy and integrity, empathy and integrity.* I thank him and take the word-gift and store it away carefully. It is on the tip of my tongue to ask him what a zodiac of the land is, but instead I decide to buy a copy of the book and find out for myself. I turn to leave, and on the spur of the moment, and apropos nothing much I tell him I will be travelling to Iona in Scotland in a few months. 'Then you need to get in touch with Anthony Thorley.

In fact, he might even be there when you go up. He works with landscape energies,' says Yuri mysteriously.

I file this information away too and feel a little thrill at what the day has yielded. I've not done badly, I think as I head outside into the sunshine, but I need to find out more about this zodiac of the land.

On the train back to London, I read a few chapters of the book but I struggle with it, can't get my head round it. Maybe it's too arcane for my blood. Is the Glastonbury zodiac a genuine map of the stars mysteriously mirrored by features in the land? Or is it, as one sceptical journalist puts it, 'a subjective interpretation of field boundaries and streams'?

I return to the book and come to a chapter written by Anthony Thorley. My head starts to spin. He writes about what he calls effigies in the land, corresponding to the zodiac signs. It's bizarre to say the least. 'In addition to the outline of the effigy... are all the aspects of history, folklore, legends and modern happenings which are found to relate to each zodiacal sign in an extraordinarily rich and apposite way,' he writes. 'Psychological projection can logically explain the outline of an effigy, even two or three – but thirteen all in the right place on the map is more problematic.' I'd thought there were 12 signs but no, there is a 13th, mysteriously called the 'Girt Dog of Langport'. I am completely bewildered.

But I'm also too curious to let it go. And the wisp of a memory surfaces: years ago I remember being given a thin map of something called the 'Kingston zodiac'. 'Because you live in Wimbledon. Kingston's just down the road and I think you might find it interesting,' said the man who'd given it to me. I'd unfolded it and the mysterious figures and intersecting angles and sharp lines on it had made no sense to me – certainly they had had nothing in common with the zodiac I was familiar with, the one linked to the 12 signs and horoscopes that I secretly read online. But the zodiac signs are more than that, they're inspired by the constellations that are tied to how the Earth moves through the heavens.

I've never explored this Kingston zodiac and for all I know the map could still be gathering dust under a pile of books at home, but it seems this is not the first time I have come across this arcane subject.

* * *

'No, I don't know anything about the Glastonbury zodiac,' says Will when I ring him from London to thank him for introducing me to Heloise. 'But I know someone who does,' he adds helpfully, pulling another rabbit out of the hat. He gives me an email for a man called John Wadsworth. 'He's a local astrologer,' he says.

John is a bearded and pony-tailed graduate with an MA from Bath University in Cultural Astronomy – I didn't know you could study such a thing – and a fan of Jungian psychology. I drop him a line. 'It would be great to learn a bit more and to go for a walk with you and see the land through your eyes,' I say, explaining a bit about what I've been up to.

'Happy to do that,' he replies, obliging and helpful. And then, bafflingly: 'I live on the Libra figure, quite close to the centre of the zodiac.'

Our walk, he tells me, will be a taster of a two-day tour of the Glastonbury zodiac that he leads in the countryside around the town. Coincidentally, he leads the walks with Anthony Thorley. (I begin to see that it's a small world, the world of landscape zodiac enthusiasts.) It's a sort of ritualised walk. I'm slightly awed that such a tour exists, and genuinely sorry that I can't make the full weekend. Then I giggle, and think that if my birder and conservationist friends knew what I was up to, they'd be rolling their eyes.

A month later, I'm back in the land of Avalon in a cafe off the main drag, almost but not quite opposite Glastonbury Abbey. I'm peering over a delicate, faded map that John has unfurled. He is pointing out figures associated with the zodiac signs, superimposed over the land. There's a Sagittarian archer, a Scorpio scorpion, an Aries ram, a Libran dove, and the rest. 'A visionary named Katherine Maltwood was the first to discern a zodiac in the land,' he says. 'She called it a "Temple of the Stars". I believe it lay latent in the landscape until she discovered it.' He also believes that it is an otherworldly *genius loci*, the spirit of the place that had ignited her passion, and her discovery.

John tries hard to explain it to me, but he may as well be speaking Klingon. It would help if I understood the zodiac signs and their symbolism and the various constellations, in a more sophisticated way than I actually do. He says that there are specific, modern-day features in the landscape that, for no logical reason, seem to correspond to this zodiac. Synchronicities abound. In the area associated with the Aries, for example, there are lots of modern-day pubs with 'Ram' in the name, and a string of kebab shops, sheep fields and hills and bumps in the land that when traced create the figure of a 'paschal lamb' – a lamb sacrificed at Passover. It sounds bonkers to me.

Even though I have a deep respect and fascination and empathy for other ways of seeing and experiencing the land, I have to ask him – nicely – if it is maybe possible that he is just seeing things and projecting meaning on to them. You'd have to be spectacularly open-minded not to think so. I also can't believe that I am now playing devil's advocate. Glastonbury and its eccentricities has made me – *me*, with all of my own alternative leanings – sound like the rational-minded sceptic. Oh, the irony. But John is not remotely offended. He merely shakes his head. Maybe he's heard it all before. 'No, there are too many coincidences,' he says.

I get that when you experience any kind of synchronicity, it can feel like the tectonic plates are shifting and you glimpse a reality you want more of, even if you can't explain it in a logical way. But all of this is going way above my head.

'To even make sense of any of this you have to be on board with the idea of the imagination as a collective field, rather than something that just resides in our heads. The spirit of the land, the *genius loci* is part of that collective *imaginal* world,' he says. 'Katherine Maltwood's discovery of the zodiac arose through the collective field of her imagination, but not in her mind per se.'

Does he mean this imaginal world is a sort of in-between place, a liminal space? And not made-up as we often interpret the word 'imagination' to mean? He nods.

If the land is actually consciously, physically manifesting features linked to the zodiac and the constellations, through this imaginal realm, then one might say it is a very playful *genius loci*, creating a quirky riddle and mystery for us humans. But why? To spark connection, a creative response, a flowering of expression? I barely understand the questions I am asking, never mind hoping for an answer that will make sense to me. I try to shield my confusion by peering into my mug of hot chocolate.

John looks up from the map. 'Look, it doesn't matter if you don't get it,' he says. 'None of this fits easily into the boxes that currently exist. What does matter is that the walks I lead are offering people the enchantment they are – we are all – hungering for. It's a journey of imaginative self-discovery. The land is a playground for the imagination,' he says, emphasising again that imagination doesn't mean 'made up' but something bigger.

Well, the walks are in a lovely part of Somerset. There is that, and it's the one thing I *can* get a handle on. In the car, we head south through more of the deliciously gentle countryside. We sweep past meadows and orchards, and hedgerows, tiny villages, the sky electric blue overhead, the sun full tilt. We turn this way and that down narrow lanes, drive through random village streets, park the car here and there and pop our heads into miniscule churches. John points out symbols that he says correspond to the Glastonbury zodiac. Later, I check my notes and see that I have scribbled down odd words with big, frustrating question marks around them – 'Mary Magdalene? A pelican? St David, patron saint of doves? Libran Dove? … Something about a church pew I don't understand?' – but at the time I nod sagely. I stop asking questions because I know I won't understand the answers even if I am hand-fed them. And because I don't want to drive my guide mad with frustration.

'Let's park here and go for a walk,' says John. Thank God, I think. I am desperate to stretch my legs. I have had enough of church interiors. We cross a field in which sits a rather defiant-looking yew. There's a faintly forbidding wooden gate up ahead. We walk through and down a slope straight into a muddy bog. I nearly fall into it, but John pulls me up as my boot lets out a giant squelch. We soon reach a trail, which runs alongside the River Brue. 'It's the Scorpio tail,' he remarks. This, he reminds me, is the area where Katherine Maltwood originally mapped the Scorpio zodiac symbol.

It feels murky and claustrophobic down here, with overhanging trees, bayou-like. 'It's like entering the Underworld, the River Styx,' my guide says. It's true the river looks stagnant and claggy and dark. 'All of this corresponds to the archetype of Scorpio,' he says. I'm grateful when we emerge from the wood and into the sunlight.

We drive some more, stop at the side of road along some fields and enter into another wood. 'This is Withial Combe,' says John. 'Glastonbury Festival is held not far from here, on the other side of Pennard Hill.' The woods here feel lighter and brighter – 'we're in the hind quarters of Sagittarius,' says John, but I've tuned out, had enough of the zodiac stuff. I might read the horoscopes but I am not, I have decided, a zodiac enthusiast. This is a type of dialogue between land and human that is beyond me. I lack the persistence, the commitment, the curiosity. I'm a lightweight. In the Glastonbury zodiac anthology, Anthony Thorley had written: 'Casual or diffident contact will not gain us entry and the resultant experience will be sparse and superficial.' I think he may have been talking about me.

Eventually, we reach a waterfall and the sun pours in. We stand for a long time, quietly, each of us lost in thought. I am frustrated by this particular landscape mystery or 'theatre' and my inability to even feel some sliver of connection to it. But then, I think, does it really matter? I am standing in front of an exquisite waterfall, bathed in sunlight. Maybe the *genius loci* of the land has led me to this: maybe it knows my limitations. Maybe this is enough.

Back in London I get on the phone with Anthony. He is a former psychiatrist, and so enraptured by the study of landscape energies that he's managed to weave them into his PhD. For Anthony, the idea of dialoguing with the land is entirely natural. 'Use the analogy of gardening,' he says when I tell him I'm not too sure how to talk about this with other people – not surprising given my own shaky grip on it. 'It's very British. People get gardening. People are in dialogue with their plants, they talk to their plants, do they not?'

His wife, Celia, has created a kind of outdoor temple in the vast garden of their home near Bath. It started off as a wild patch of woodland. 'It was co-created, with the garden,' she will later tell me when I visit. Walking along the winding path of the garden is an enchanting experience. It is a profusion of seasonal colours and hues, wildflowers and shrubs, overlooked by very tall, mature trees. But there is a wildness here too. 'When I tried to grow or plant things in a way the garden didn't like or want, they just didn't take and I'd have to try again,' says Celia.

Conspiratorially, she shares with me stories of those who've walked in the garden, stories that involve the appearance of birds and butterflies and presences at auspicious moments. Anthony – who, in the flesh, in his battered Panama resembles a character in a Graham Greene novel – suggests I talk about such things lightly. No sledgehammer. 'Be lightly self-mocking.'

If I laugh at me, others will laugh with me, not at me. To be honest, I don't hold out too much hope of that happening. And yet, the premise that the land is alive, that it has a spirit – what is so strange about that?

# A Temple in the Land

The autumn equinox is approaching and the winds are shifting. I think maybe, just maybe, I am learning to listen. A tiny bit. I might be on a walk and there might be a stiff headwind turning me away from the path I was going to take and onto another that turns out to be three times as lovely. There might be a bird of prey hovering above, beaming goodwill. I might get lost and when I am about to give up or turn back or just stay lost, I find myself funnelled into the right place at the right time. I have to stand still, stay silent and listen very carefully. And even then my head might tell me that I'm not hearing anything at all. It's a subtle art, this listening business.

So why are there not entire classes, clubs, schools devoted to this most delectable pleasure, this art of deep listening? There is no more delicious feeling than when the mysterious force behind the nature out there lets you know of its presence. It may not make logical sense but it feels better than anything, sometimes even better than sex.

Maybe this is what death feels like. Maybe in death – when you become pure consciousness and are everywhere at once, as some who have come back from the brink say – this is how whatever is left of you feels. I once interviewed a woman who had had a near-death experience and four years on she was still on a high, glowing like an incandescent alien being. So great was her glow that it spread to me and I too was on a high for weeks after meeting her. I wish I had asked her if she spent time outdoors, if she went for walks in wild places and if so, how did it make her feel?

I have set my heart on travelling to the tiny Hebridean island of Iona. What has drawn the hermits, the saints, the mystics and the monks here? Can I connect with whatever it might be, in an intimate and personal way? I've read travel features, grazed on the island's history, read up on St Columba who started a monastic community here in AD 563 – a time so long ago I laugh out loud when I read it. The saint converted non-believers to the faith and the island became a place of Christian pilgrimage. What drew the believers and those who came before them to the island? I'm curious.

I am also going to put my money where my mouth is and conduct a little experiment. I'm not going to plan my trip. Instead, I'm going to ask the land to guide me and then I am going to listen and do as I am told.

'So basically you're going to rock up and then be a bit spontaneous?' asks one of my more sceptical travel-writer friends. Yes and no. Because whatever direction the land nudges me in is the direction I will take. Subtle difference. And no one just 'rocks up' to Iona unless you happen to live on neighbouring Mull. It's a long journey – 12 hours from London when you include the flight, the bus, the train, the ferry, another bus and a final ferry. A marathon journey, and then to get to the island devoid of any plan, much as I love spontaneity, seems a bit much. And yet, why not play a little, be a child? How often do we get to plot – or rather not plot – an adventure in this way? Not often enough in my case.

Iona has umpteen dazzling white-sand beaches with waters so clear and turquoise they look as if they belong in the Caribbean. It has serpentine stones that are a mottled gold green with a strange snakeskin-like pattern – the name is said to come from its colouring. The island is formed of rocks that are a mind-boggling 200 million years old. It has wild corners and slippery headlands where you can get lost, trip and fall. I read that Iona exudes a haunting, otherworldly peace that cradles you and eases your cares. It is a place where the veil between worlds is thin.

I confess, it's a photograph of one of the beaches, the White Strand of the Monks, that first tempts me. It looks romantic and inviting, but then I read that monks were murdered on this very beach by marauding Vikings. Not so romantic. But I still want to see this beach for myself, and so I book all the tickets and work out the logistics, which takes for ever. It takes commitment to get to Iona, that much I know. Then Hurricane Irma hits over in the States and the effects can be felt all the way across the North Atlantic on the tiny island. I'm in for a rough ride, weather-wise at least.

* * *

Three weeks later, I'm sitting on a bench on Glasgow's Queen Street Square, waiting for the train to take me to Oban. This is the second leg of my journey, or third if you count the bus from the airport into the city. The autumn sunshine is delicious and I am inhaling lungfuls of air – all that canned travel and I'm starved of it. I'm also contemplating the slightly cryptic message I've received just this minute from Anthony Thorley, the landscape-energies expert. Coincidentally he has been on Iona this week but he left yesterday. I'd emailed him to tell him of my non-plan and was disappointed to read we would miss each other, but then came his email:

*There is a diagonal valley rarely frequented called the Glen of the Temple. Walk down there and contemplate What Temple? Where, on what plane is this temple?*

Aha! This is it. This is the land reaching out to me, nudging me. I will look for this temple. I'm not sure what it is exactly, and I know nothing as yet of the story behind it, but this is amazing. A sign. I carry on reading:

*Find out about the incredible geology of Iona… how it has some of the oldest rocks in the world. This must have been an energetic attraction for all the sages in the past.*

Interesting. Even Anthony's phrasing – the vaguely seductive, fizzing 'energetic attraction' – appeals to me. I leave the park bench and the sunshine in a semi-trance. I shuffle into the station and don't mind that the acoustics in the station are terrible, amplifying the clamour, or that the queue to board the train is more like a Delhi-style scrum. As soon as I've found my seat – a window seat, thankfully, for I've heard the views up to Oban are spectacular – I google 'Iona geology' on my phone, praying the signal won't cut out too soon.

I learn that Iona is made up of ancient rock, granite gneiss that is nearly three billion years old. It's the oldest rock on earth, so the island is not just a pilgrim's paradise but a geologist's one too. People who live there and people who visit properly – who take the time to get off the beaten track and explore its wilder corners – speak of its special energy. It is something to do with age and geology and wildness and place and spirit coming together in a magnificent alchemy to create a kind of magnetic, attractor 'scent', which has pulled people in over the ages.

I take notes about the various island features: Dun-I, the island's highest point; a circle of stones known as the Hermit's Cell, where St Columba was believed to have meditated; the quixotic-sounding 'Well of Eternal Youth' (I particularly like that one); the beaches, of which there are so many – Port Ban, Traigh Mor, the Bay at the Back of the Ocean, Colomba's Bay with its serpentine stones…

My phone dies, and as the train speeds onward to Oban and the landscape closes in, my mood shifts. We whizz past pine forests, hills, lochs, heather-covered slopes and mountains, a convergence of rolling mist and autumnal golds and a patchwork of green and dark shadows. The sun disappears, replaced by a misty, gloomy veil through which rays occasionally burst like the high note in a long orchestral work, creating odd shadows on the land.

I once spent a few days on Shuna, a Hebridean island with a permanent population of two. Surrounded by the water, I'd felt at ease, safe and buoyant. But this mountain and loch scenery, without the escape that the sea offers, feels oppressive, masculine, impenetrable. If these hills could talk, they'd be hard and unyielding, not giving away much. The observation is unfair, I know, as I haven't done any actual walking in them. Still it seems to me that a rigour, a faintly dour single-mindedness is required if you're going to walk these wild hills. In short, the scenery leaves me feeling wanting and inadequate and not very British at all.

On the coach across Mull, I get talking to a woman who is heading to a retreat on Iona. 'It's called Paneurhythmy,' she says. 'The word means "the movements and rhythms of the universe". You dance outside early in the morning. It's a kind of prayer with your body. It's meant to heal both yourself and the land.'

'Sounds… interesting,' I say, happy to have someone pleasant sitting next to me. But the prospect of dancing on a beach leaves me cold. I once did just that on a beach in Dorset and hated it so much I cut and ran. Doing arabesques and making strange shapes and curling on the rocks like a foetus, and peering into rock pools with my bottom sticking up in the air with others doing the same while passers-by giggled, wasn't my thing. I chat some more to the woman on the coach and learn that the woman leading the Paneurhythmy retreat is married to the man who first discovered these landscape temples, including the one I'm trying to find. Small world. His name is Peter Dawkins and he is currently on Iona. Anthony Thorley had mentioned him to me briefly. Maybe I can enlist his help? I make another little note.

It's sunset by the time I cross over from Fionnphort and arrive in Iona's tiny harbour. I could have made it to Australia in less time, I think. But walking down the gangplank, I'm taken aback by the halo of soft golden light. The luminosity feels like a hug. And there is the quiet too, which is nothing like a city quiet with its eternal hum. No, within this quiet are oxygen, vitamins, sustenance, the gentle lapping of waves. A good silence for listening. And like that, in an instant, the interminable journey has been wiped clean from my memory banks.

A woman waves at me from the end of the pier. Sara runs the farmhouse B&B I'm staying at. She's a pretty, hardy, but not unfashionable islander. In her blouse and jeans, she looks a lot smarter than I do with all the outdoors kit I have on and which makes me feel faintly ridiculous. When I go anywhere rural I never see the locals dressed this way; they just pull on some old mothballed jumper and some kind of mysterious wellie–walking boot hybrid. Why do we do it, I think, not for the first time lamenting all the finicky loops and tassels on my rucksack, the purple fleece top, the blue waterproof, which are all like a giant sign screaming: 'Londoner on the loose!' On the other hand if I *don't* wear the gear, I feel the eyes on me thinking, even if it's a fleeting, barely conscious thought: 'Oh, Asian woman. New to walking. A novice. Out of her depth.' I don't want to be judged this way and I'm insecure enough to care.

'All the day trippers have left now, so now you'll have the island to yourself,' says Sara as we drive the half-mile on a silent, empty road to the farmhouse where I am staying, west of the pier. I have an impression of fields and sheep and a big sky, the occasional cottage. The day trippers are needed to keep Iona afloat, but for the locals they are a mixed blessing: a big conga snake that heads to the village, to the abbey and then slithers back onto the ferry. They will never understand the everydayness of island life, the agricultural life, the isolated life. They will never know the land intimately, as a friend. They will only see the island through the lens of specialness, in season, and maybe they won't go any further than the village. I am not so different from a day tripper, much as I like to think of them, sniffily, as mere 'tourists'. Then again, I am staying for nearly a week and I *am* looking for an invisible temple...But who am I kidding? To a local, I'm as much a tourist as the daytrippers. Traveller, seeker, vagabond, explorer, adventurer: these are the names we like to give ourselves, because the word 'tourist' with its connotations of consumerism and superficiality makes us cringe. But is a pilgrim a tourist? For a pilgrim, you might say the destination lights the journey and for a tourist, the journey begins *at* the destination. Or does it? I'm tying myself up in knots, and I don't think the island cares much how I choose to define myself.

In the house, Sara goes through the drill very fast: 'breakfast... living room... need anything, ring a bell.' She is eager to enjoy her evening. 'Got to get to the pub for dinner!' she says. For some reason it surprises me that out here a woman whose husband keeps sheep cares about fashion and has a social life. And that there is a pub. Will I be brave enough to enter, though? I feel a tiny kernel of anxiety. How will drinkers on a tiny island receive a brown face? But it's a holy island so maybe they'll be kind and tolerant?

In my room, I face a field of sheep, a white pony named Dio, and the sea. The waves are flowing fast now, pushed by the wind, as if they are in a hurry to get somewhere. The scene is perfect. But the quiet after the long journey, the lack of anyone to unwind with, is oddly unsettling. I'd do anything for a drink. But the thought of stepping into the pub by the harbour on my own, and the inevitable stare – even if it is discreet – is more than I can face tonight. In tiny, isolated places I often wish I could pull on a cloak of invisibility.

Instead I walk down to the harbour, soothed by the waves and the dipping sun, make a phone call to my mother, and then turn back up the hill and crawl into bed with oatcakes and miso soup and apples. It will have to do for now.

⋆ ⋆ ⋆

I groan when the alarm goes. Nothing worse than having to rise early on a grim day, and today is grim. I can hear the sheer ferocity of the storm. Today is not the day to be searching for the Glen of the Temple. 'Postponed till tomorrow' is the cop-out mental Post-It note on my forehead. I'm a wimp in the rain, and it is lashing down. There's hail too, bouncing off the window. When I crawl out of bed, freezing cold, I take a peek outside and see that the sea is a froth of white peaks. I've not seen a single photo of Iona in the rain and now I know why. The landscape suddenly feels alien and harsh to me, not like a landscape that wants to meet me. It is all fine and well wanting to connect with some indefinable yet very real spirit of a place – real to me anyway – when infinite blue and dazzling sun are taking up residence in my mind's eye. But how can I do this in gale-force winds and rain hurling in all directions?

I am not comfortable with this wild, untamed, gusting ferocity, and don't know how to engage with it. The blood in my veins feels hot, descended from a hot-blooded, passionate people. The cold, the wet turns my blood sluggish. And I shirk from darkness, literally and figuratively unless we're talking about something like star-gazing or walking under a full moon. My parents struggled against the darkness that infiltrated the land they were born in, the darkness of apartheid. They struggled with that darkness and by some miracle and by dint of my father's will they emerged into the bright light of a winter's day in Canada. So is it any wonder I don't like to dwell on or in darkness? That I have no need for any allegiance to the tumultuous side of nature because the tumultuous defined my parents' early years? I don't need to embrace the dull misery of a trudge in the rain. I know misery through the stories I have grown up with, the misery of alienation and injustice. But I am here and I need to go outside.

First: breakfast. Over the full Scottish with farm-fresh everything, I attempt conversation with a mousy, forty-something teacher from the mainland. 'I take a few days off now and again to walk,' she says when I ask her what brings her here. 'I like it here.'

'You don't mind the rain?'

'No, this is Scotland!'

I kit up in my room and the minute I step outside, rain batters my face and the wind pushes me back. I trudge glumly past Dio, whose back is turned to me, his white mane flattened, and past the sheep, stoic and miserable. I hesitate on the road. Do I turn west towards the dune grassland of the Machair and maybe take a south fork towards St Columba's Bay at the southern tip of the island? Or is that insane on a day like today? Or do I make for the Bay at the Back of the Ocean, a poetic name for one of a series of beaches beyond an incongruous golf course? Or do I retrace my steps from the night before and turn east towards the village and carry on beyond it to the abbey and the north of the island, where I'll find Traigh Ban, aka the White Strand of the Monks, where the massacre happened all those centuries ago? I stand indecisively for a good full minute. Which way am I pulled? It's hard to listen in the storm.

I decide to veer away from people and the village, and trudge out to the Machair past small farms along one of the roads that cut across the island. I can't see much with my hood pulled low and I am soon engulfed in the storm. It's crazy weather for a walk, I think, squeezing water from my useless cloth gloves. To think there is a heatwave going on in London right now! I cross the golf course, wondering who on earth comes to Iona to play golf, and then follow a dirt track down to the beach. Here I take a sharp intake of breath.

How many ways are there to talk up a beach? Today I try 'a curved expanse hugging a sea that is a startling aquamarine'. It's not an especially poetic description but it is true. It's not the inky black grey I'd have expected either. The foreshore is strewn with seaweed and carpeted with hypnotically beautiful beach pebbles: red ones speckled with black or bits of marbling, terracotta with green veins, vivid ochre stones, smooth grey ones. Some make me think of the glittering skin of a snake, a creature that is also hypnotically beautiful – that is if you aren't terrified of snakes, which I am.

There's some action out of the corner of my eye, and I look up to see a flock of Canada geese flying towards the beach in a 'v' formation before making a slow turn and wheeling off (somewhere warmer, I imagine). Where have they come from? Iceland? Greenland? Seeing them always comforts me and reminds me of my other home, far across the Atlantic. In their wake is a strip of violet-blue sky, and within moments there are whole patches of the blue amid the heavy clouds, bearing light and the suggestion of sunshine. And then the rain lets up. It is as if the geese swept past and washed it away. Good. I've no idea how long it will last so I had better make the most of it.

Like a beachcombing Alice in Wonderland, I walk along the beach, slip-sliding over the pebbles, before moving up to higher ground, and down again along the sandy trail, through gnarly outcrops, till I come to Port Ban. The cove is wedged in between the rocks. Here the tide is in, the sea is turquoise and the foreshore is made of small, silvery, dusty-pinkish shells nestled on a bed of crystals, tiny snake teeth and stardust – how have all these tiny crushed shells come to be? It's an entire world of shells in miniature. I've found myself a honeypot. It's stopped raining. I have the beach to myself. I have a flask of hot tea and some chocolate. I'm busy congratulating myself and feeling the particular joy that comes from creating a moment, when it starts to rain again. Not just rain but howling rain. And wind.

Any sane person would now head back to the warmth and shelter of their lodgings and dry off. But I am not any sane person and I am so fired up by having this stardust beach to myself that I decided to carry on walking. My plan is to hug the coast – when in doubt, hug the coast – but that becomes impossible once the outcrops and hills become more hills and more rock and no tantalising dips to white-sand beaches. So I go inland and lose my sense of direction. It is desolate and eerie in this land of moor and bog, a place where shadowy presences seem to lurk and the rain falls like sheets. It feels like a place that prophets and saints might hide away in. The next day I'll meet a man who tells me this stretch of the island is known as the Great Loneliness, which in hindsight makes perfect sense. The Great Loneliness is not enjoying my being here today, judging by how I keep being nearly knocked off my feet.

I know that somewhere in my general line of sight is Dun-I, the island's highest point, and somewhere around it the Hermit's Cell, but there are two hills to climb up and over before I can even get to the base. Most keen walkers who visit Iona walk up to the top of Dun-I, but doing it on a rainy, slippery day when I am alone is probably quite a stupid idea.

The tempo of the wind, the hail and the rain suddenly go up a notch. It's as if some mysterious conductor in the sky has gone a bit mad, waved his baton wildly and shouted to his orchestra: 'more, more, more!' I am fighting to stay upright. If I'm not careful I'll end up a casualty, so I rein in my curiosity and retrace my steps. Back, and back, and back I go, across the Machair, the golf course, the road.

Something in me has relaxed by the time I reach the farmhouse. Every trace of gloom has left me and I am wide-eyed and energised.

⋆ ⋆ ⋆

'Can you tell me where the Glen of the Temple is?' I ask a grey-haired lady serving at the counter of the big community shop in the village after lunch. I'm full of hot soup from a self-service cafe, one of about three places where a person on a budget can eat on the island. The others are smart hotels, filled with American day trippers. Americans, I deduce from the accents I've heard, love Iona, despite the fact that owing to the stormy weather, all the ferries back to Mull have been cancelled.

'Never heard of it,' she says curtly, turning away to serve a customer. Aside from my host, Sara, this is my first ever exchange with a local on the island. I feel the slight, not least because there is a big 'Welcome' sign outside the shop. Some welcome. I am pissed off, I am hurt, I am secretly wondering if she is a tiny bit racist, and I am also wondering if this Glen of the Temple actually exists. I mean, this woman is an islander. Surely she'd know?

Then again, I have only asked one person and already I am doubting my mission. The issue is trust. If I want the land to guide me, I need to show trust and commitment. I need to treat trust like a newborn, to cradle and nurture and treasure it. The island, I suspect, is not going to offer me a fast fix. It is an ancient island and it is not going to yield its secrets easily. All manner of pilgrims and shepherds and monks and nuns and hermits and visitors have come before me and stayed for months. I am just one in a long stream of seekers and I need to be patient.

By the time I have walked to Traigh Ban on the other side of the island, I've walked off my anger at the woman in the not-very-community shop. The beach may have been the site of a massacre, but it bears none of the scars: no atmosphere, no strange energy or depressing, intangible fug. The fickle sky may have something to do with it: it has cleared once more, the sand is a hallucinatory white, the sea turquoise, and not far from the shore a sandbar shimmers. If anything it is more exquisite than Port Ban. I can see the hills on Mull in the distance. The moment I get settled on a nice rock, unscrew my thermos and start drinking my tea, the rain begins to fall.

I've had enough. I throw in the towel and turn back. Iona is playing games with me.

* * *

Having got a polite brush-off email from Peter Dawkins (the man who the woman on the bus told me had discovered Glen of the Temple), I get a brainwave the next morning and walk down to the community notice-board on the pier and ring up a guide whose phone number is listed on it.

'Hello, Jana. Have you heard of the Glen of the Temple?' I ask in my most persuasively friendly voice. Or is it the Temple of the Glen? I can't quite remember. But I am feeling more confident this morning. The sky is a clear, bright blue. And last night I ate dinner alone in St Martyrs, a restaurant on the pier in a room adjoining the pub. I drank a glass of wine to give me courage. It worked: I felt a rosy glow, and the seafood on my plate was tasty, and I didn't mind that I was the only woman – and the only dark-skinned person – sitting alone, and then when I got back to my B&B I slept like a log.

Jana wants to help, I can tell. She is nothing like the curmudgeonly woman in the not-very-community shop. 'I haven't, but let me ask my husband,' she replies in a friendly way. I can hear her calling out to him in the background.

'Not heard of it,' he says. My heart sinks. Then, there's a long pause and... 'He's found it!'

It turns out her husband has found Glen Temple on Google Earth. Inwardly I roll my eyes. There's nothing like technology to tarnish a mystery. But still, the information is helpful. 'It looks like it's inland, sort of west of the abbey, somewhere in the centre-ish of the island,' says Jana. 'Try talking to Daniela. She works at the Iona Heritage Centre. She might know more. It's her kind of thing.'

I thank Jana and call Daniela. She answers on the first ring. 'You want to do all this research and you're only here for three days?' she says, sounding unimpressed.

'Um, yes, that's the plan.'

'OK, well, if that's what you want. I'm working in a hotel on the island today' – she has a few jobs, apparently – 'so go and kill some time. Have your little adventures,' she says a tad patronisingly, 'and then come and see me.'

I decide I am not going to wait for Daniela, but will go to the Heritage Centre anyway. It is a pleasant walk past the ruins of a medieval nunnery, a school and a pretty parish church. It has a tiny museum in it, filled with the paraphernalia of island life. There is a shop too, with serpentine necklaces on display and, round the corner, a cafe. The older woman at the reception desk of the Heritage Centre is kindly and soft spoken. Best of all, she has heard of the Temple of the Glen.

She unfolds a map of the island, far more detailed than the one I have, which I'd picked up on the ferry to Mull.

'Here it is,' she says, pointing to a valley in the middle of the island. To get to it, you have to go through a passage over the rocky hills and bog west of the village. 'It's been a very long time since I've walked there,' she says in a soft burr. She tries to recall where to start from the village, but can't and says that in any case the bogginess and the lack of footpath will make it a difficult walk. There looks to be another far easier route, via the Machair, but she mentions farm fences that I won't be able to get past.

I could do a recce of either, but how will I know when I've reached the Temple of the Glen? There won't be a sign, declaring 'mystery here'. For all I know, I might walk right on past it. In my mind's eye I picture some sort of shimmering gold pagoda. The truth is I am not skilled enough in the art of deep listening to know it when I come to it. And as yet I have no context for the Temple.

There's a house owned by the Findhorn Community near Traigh Ban beach, the woman tells me. 'You can ask the custodians there. They ought to know more – and something of the story behind the Glen too,' she says.

This is good news – a good omen. I feel a rosy glow of nostalgia when I hear the word 'Findhorn'. It's an alternative eco-village near the Firth of Moray and co-creating with nature is at the heart of everything they do. I once stayed there for a week, at a conference where radical economists, alternative educators, artists, farmers, permaculture practitioners, human rights lawyers, indigenous spokespeople, animal rights activists, adventurers, retreat leaders, and even a national newspaper editor mingled and partied. People spoke, listened, argued, shared stories and some even bonded, naked, in the hot tub at night under the stars. Drunk after a party – the parties were great that week – I snogged one of the speakers, overslept and nearly missed the train back to London.

I'd been invited to cover the event for a magazine and on the way up on the train, I'd only been mildly curious, a little blasé even. But that feeling had given way to something deeper and more powerful as the week wore on: the knowledge that I wasn't alone in wanting to find my way to an inner relationship with the natural world and the landscapes I was moving through. To call in at the Findhorn house on Iona, to my mind, would be like visiting family. Or so I imagine.

Anyway, the sun is delicious and even if I get lost – the directions the woman gave me are of the 'past that farmhouse, down that gate' variety – I'll be close to Traigh Ban beach so going there seems like a good next step to take.

Following the directions as best I can, I walk through a gate and down past a farmhouse while hugging a fence line. I immediately spot the Findhorn house. It is close to the sea, surrounded by a rambling garden and a low wall, with a little gate on the side. It is friendly-feeling but when I rap on the door, a man with a pinched, Fagin-like face – straight out of a Dickensian novel or a Grimm fairy tale – fairly flings it open.

'I was in the middle of making a fire,' he rasps, standing sideways to the door as if only half of him is willing to acknowledge my presence. What is it about this island and its welcoming committee? 'You've interrupted me. The smoke detector might go off if I stand here any longer,' he says.

I apologise for turning up unannounced, and tell him I once stayed at Findhorn and have been sent here by the woman at the Heritage Centre so that he knows I am not some random interloper. 'I'm eager to find the Glen of the Temple. Have you heard of it?' I trot out the line for the fifth time in two days.

'Wait here,' he says, turning back into the gloom of the cottage before returning a minute later and thrusting a giant blue binder in my arms. 'Read this while I check on the fire.'

I step into a little anteroom that overlooks the garden, and start flicking through the binder's pages. In it are notes, in the form of an extended email from someone called 'Sverre' to 'Niels'. Sverre talks about the geography and geology of the island and its history. The glens are described as magical 'fairylands of lush grass'. The 'fragrance of sweet grass, lavender and sea are an intoxicating aromatic cocktail,' he writes. To him, Iona is far more than the sum of its parts. It is an ancient land, alive and teeming with magic, no less than a 'sacred portal that channels a stream of radiant, divine energy', a 'harmonic blend of sea and earth', the equivalent of the mystical 'Third Eye' that sits in the middle of the forehead, the gateway to a higher consciousness and insight. Even St Columba recognised its special, sacred aura, he writes to his friend. This is more like it.

There is more in this vein and about Staffa Island too. About an hour's boat trip away, it is uninhabited apart from the seabirds and is described as a temple and a 'pulsing generator of white crystalline energy'. But what of the Glen of the Temple? Or the Temple of the Glen?

I hurriedly flick through a few more pages. Iona is composed of ancient gneiss rock, one of the oldest on the earth – I remember reading this on the train up to Oban. There follows something geology-related about how the rock, over time, has turned into a more crystalline structure that emits an electromagnetic field.

'All crystalline structures have an electromagnetic field,' writes Sverre. 'The greater the density of the structure the stronger the field.' The peace that blankets the island is down to an electromagnetic field then, partly created by its gneiss rock and some interplay between it and Staffa. On Iona, there is also some sort of anomaly created by 'strata upthrust' that results in a magnetic release that 'thins the veil between planes'. Science, he goes on to say, has documented the relationship between magnetic anomalies and psychic phenomena and experiences.

I peer into the gloomy hallway and see that the man is now fiddling about with a pair of bellows, having rearranged logs in the fireplace. It's not that cold out, I think. I keep reading.

Sverre writes that the metamorphic nature of the gneiss – rock that has changed due to extreme pressure and heat – coupled with its magnetic fields has allowed Iona to act as a kind of receptor for energetic frequencies that are projected from Staffa, which is made of basalt and other rock types. These energetic frequencies find their way into pockets around the island that are a match for the Staffa frequencies. These pockets correspond with pure musical notes and the chakras – those spinning energy centres that are more often associated with the human body but here relate to the body of the island. In other words, the island has chakras, and these chakras are considered places of potent energy – temples in the land. It's a lot to get my head round.

I hurriedly run my eyes down the page: it seems there are six of these chakras on the island, all with names: 'Angel Hill', 'Dun-I', 'Columba Bay', 'Back to the Island Cairn', the 'Hill of the Lambs', 'Signal Hill'. These are the temples of the land, thin places, places of maximum peace and serenity. In Iona, Sverre declares, there is no duality. 'The darkness cannot find a pocket to dwell'. Because of its remoteness, its sparse population and connection to Staffa, it is able to maintain the purest frequency. The island is unique.

Perhaps if I'd had more time or if I had been able to talk to Peter Dawkins or to Anthony Thorley, or if the man in this Findhorn house was more forthcoming, I might have understood this esoteric phenomenon in a more sophisticated way. Maybe I'd have grasped subtle nuances and distinctions and be privy to more recent discoveries – I have no idea when any of this was written.

If I were to delve deeper, I might hear opposing voices too. But honestly, I don't care. This information has fallen into my lap, which is amazing in itself, and it is enough for me. I don't understand geology or the mechanics of a land temple, but the poetry of it, the theatre, the mysticism, the spell it casts – this all stirs something deep within me.

Visitors to Iona are always remarking on the special peace here regardless of their beliefs. And one person's 'energetic hot spot' is another's tranquillity. The language may differ depending on your culture and belief system. But the feeling will be the same. Feeling is universal. But – and it's a big but – not one of the island's hot spots goes by the name of Glen of the Temple.

Still, I might need this information so, like the spy I am not, I whip out my phone and begin quickly snapping photos of the relevant pages. Silently I thank Sverre, whoever he is. I get what I need and then hear footsteps and shove my phone back into my rucksack.

To my surprise the man, his fingers slightly blackened from the fire, returns with a map. 'The Glen of the Temple is on here,' he says, unfolding it and leaning it against the wall. Like the woman in the Heritage Centre, he jabs at a spot in the middle of the island. Signal Hill, one of the island's energy centres that Sverre mentions in his email is mentioned on the map and… it looks out over the Glen of the Temple. Aha!

'You won't find it, though,' he says, frowning. 'It's too muddy, too wet, wrong season.'

'What about the easier route through the Machair?'

He shakes his head. 'There are farm fences and private property. You won't be able to get through to the valley.'

Anyone would think he was trying to put me off on purpose. Maybe everyone is. This is the island's secret and I am not an islander. Or maybe he is right. Maybe it is a fruitless endeavour. Maybe I should let it go. What now? I thank him and step outside.

In the garden, another friendlier-looking custodian appears. 'Have you got what you need?' he asks gently.

'Not really,' I shake my head sadly. He seems kindly, so I share my confusion and frustration.

'The truth is, it doesn't really matter what these places are called or where they are. Just wander across the island and see what inspires you,' he suggests. 'There is endless magic here in the hills and the moors and the beaches.'

Suddenly I no longer care about finding the Temple. I am here, in Iona, a land of sublime beauty. I am free as a bird. I have no commitments while I am here. There are so many corners I have not explored, so many beaches I haven't laid eyes on. I've not made it to the wild west of St Colomba's Bay and searched for serpentine stones. I have a whole afternoon ahead of me and I decide I am going to spend it doing absolutely nothing on Traigh Ban beach and isn't that enough?

Decision made, I walk over a grassy cleft and clamber down to the beach through the dunes. The sea is a patchwork of teal and turquoise and midnight blue, the sand alabaster. The scene is exquisite. This is what love looks like, I think, as expressed by the landscape. I find a stick in the dunes and trace a big, wonky heart on the beach, then look up and see a couple walking a dog, and suddenly feel a little silly. I hurriedly erase the heart with the stick and decide to check out the next beach along.

Up a hill and down a dune and through a gate I go – just a short walk, really. The next beach doesn't seem to have a name. It is wider and wilder and windier, and deserted. I walk across and spy a neat, modest woman exuding the unmistakable aura of discreet wealth. She is perched on the rocks, and seems to be focused on some far distant inner horizon. Sara at the farm had told me this morning that Melinda Gates, the philanthropist and other half of billionaire Bill is on the island. 'She is here on a silent retreat, staying at the cottage down the road. Her party fly over on a helicopter, bring all their own food, even a chef,' she'd said. Her mode of transport aside, there is something reassuring about a woman of immense power and wealth finding her way to the peace of Iona.

★ ★ ★

Seals basking on rocks, kittiwakes, cormorants in flight and shags on slippery rocks. The guide on a boat made from larch and oak is pointing out the wildlife on the way to Staffa. It is the following day and I have one more left on the island. The sky is grey, the waters choppy, the day trippers and Iona old-hands philosophical. I speak to a Dutch woman who tells me she has stayed on Iona 12 times before. She is a refreshing antidote to the frenetic humans who are permanently in a hurry, spending their lives trying to fulfil travel bucket lists. I should know: not so long ago I was one of those people.

'I can eat in the pub and no one will bother me,' she says, not realising she has given me the answer to a question I have been asking myself every time I pass the bar bit of St Martyrs on the pier. Perhaps it will be OK, like the restaurant was, and no one will stare. Once I 'do' the bar, I will have one other place left to dine on the island, the Heritage Centre cafe. I'll save that for tomorrow, I think, luxuriating in these tiny decisions.

This isn't the Dutch woman's first time on Staffa either. 'You'll see,' she says. 'It's a kind of a cathedral to nature.' Having read the emails of the mysterious Sverre and his description of the island, I have to see it for myself. Right now, I keep my eyes on the horizon while the melancholy cries of the seabirds create invisible, swirling patterns in the sky above and around us all. I think of the artist who attached tiny lights to the feet of pigeons and flew them around an audience at sunset to draw hypnotic patterns in the sky.

It is Fingal's Cave, the sea cave on Staffa with its pillars of hexagonal basalt facing the rising sun, that we are all eager to see. After about an hour of straining my eyes, peering at the horizon, I see a tiny speck in the distance. It grows and grows. Soon enough, what I see looks like this: a head of broccoli sitting on top of a giant upended xylophone, a work of surrealist art created by a mad genius of a giant. This is Staffa.

The columns of the cave are ridiculously pleasing to the eye. They are so perfect, so symmetrical, it is hard to believe they are nature's own sculpture, the result of an ancient lava flow formed some 50 million years ago. I can't see the entrance to the cave until the boat slows and draws closer, but when it does I see a gaping black hole that looks like a heart.

The wildlife love it here too: every summer a colony of puffins breeds on the island. They hang out on top of the cliffs before diving into the water for fish. I am sorry to miss them. If I am lucky, and we haven't already spotted them in the sea, we might encounter shags and cormorants, guillemots, razorbills, shearwaters, gannets, ravens, arctic terns and more. But how will I know which bird is what? I won't, but I am happy enough to think of them as a great feathered choir. In the sea, basking sharks, dolphins, grey seals, and minke and pilot whales swim about.

On the island, getting to the cave means shuffling along a narrow, slippery walkway while clinging to a rope and hugging the rock. It is slightly nerve-wracking but once I get to the entrance, I venture in as far as I can go. There the deafening roar of the sea as it comes crashing in and back out again, the collision of rock and sea, the light outside the cave and the gloom within are elemental, pure power and presence, as formidable as any deity.

The cave's high ceiling gives it a grand cathedral-like aura. *This* is a temple. Sverre had said the cave is a site of pure energy, and now I can see the crystallised basalt columns that create the electromagnetic field he had described. If I listen very hard I can hear a silence behind the deafening roar, a silence that might even be orchestrating the roar.

I read somewhere that Jules Verne had visited. He'd mentioned Fingal's Cave in *Journey to the Centre of the Earth*, as had various poets and composers, including Mendelssohn who'd written a Hebrides overture. (When I listen to it later, I can hear the rush of the water, but the music feels to me too polite, too civilised, too constrained.) Even the writer Sir Walter Scott said he had touched the 'spirit of God' here. Reading that sent a shiver up my spine. I am eager to know of the reaction of poets and musicians and artists from other cultures to this wonderful place. How might someone from the other side of the world interpret the magic here, this quality of Otherness?

I find a spot on the rocks by the entrance to the cave and let myself be dwarfed and silenced and humbled. I am in a primeval, thin place, peeking through the veil. I know just how Walter Scott felt.

★ ★ ★

By the time the boat gets back to Iona I am famished. Little do I know the island has a surprise lying in wait for me. From the pier, I walk straight over to the Heritage Centre cafe and order coffee and a toasted sandwich from the waitress with a purple streak in her hair, who looks as though she'd love to leave this island life and work far, far away. I plonk myself and my rucksack down at a table and catch myself staring at a woman across the room.

She looks familiar. Do I know her? She is grey-haired, delicate-looking, bent over her laptop. Then it comes to me: she is an artist who I'd met briefly during my stay at

Findhorn, the eco-village in Scotland. We'd had no more than a 10-minute chat at the time, but I remember her because she was the event's artist. She'd sit by the side of the stage and get busy creating as one speaker after another held forth. Her art was… unusual. She conjured spectral, abstract, mystical images by using the pressure of her fingertips on paper placed over wet paint. How odd to find her here. I decide to go over and introduce myself.

'Are you Deborah?' I ask. I notice her fingernails are long. Maybe if you do the kind of art she does they have to be.

She nods, smiling. 'Yes, I thought I recognised you. From Findhorn, right?' Nobody forgets a Findhorn encounter easily.

I am relieved and happy to find a familiar face here and I sense that it's mutual. Deborah invites me to join her. I need no encouragement and the waitress brings my food over. Deborah tells me she has just arrived on the island from Washington. She's jet-lagged and disorientated. She is teaching in Suffolk in a few weeks but had always wanted to come to Iona and so she decided to fly here first. She is staying in a house on Traigh Ban beach, she says, and is here for 10 days. 'But I'm going to Wales after the course, to see yew trees with my husband. He is going to come over and join me.'

I try to explain why I'm here. I tell her straight up about the Glen of the Temple, how I'd been so eager to find it but then had given up. Sort of. I dig out the map from my rucksack and point to the spot in the centre of the island. I feel like something of an expert now. 'Everyone's been pretty discouraging so far.'

'Well, why don't we try to find it together?' says Deborah.

And just like that, something is set in motion. I'm going back to London the next day so I think we may as well try. We leave the Heritage Centre cafe and start walking in the direction we think the Glen of the Temple might be in. Yesterday I wandered around the ruins of the nunnery. Now as we pass them, Deborah recognises a familiar face – a grey-haired woman with a short, severe haircut and a serene face, a Buddhist who tells us she lives on the island during the winter months. I have no idea how the women know each other, just that they do. The woman, whose name is Sally, is in walking gear and carrying walking poles. It seems we are walking in the same direction and so I ask where she is heading.

'A place called the Glen of the Temple,' she says.

My jaw drops and I stop in my tracks. Have I heard correctly? I'd spent four days trying to find this place. How on earth…? I give Sally the best bits of my story and she looks thoughtful.

'I often go there alone. I like the peace. It's a special kind of peace,' she says. 'We can walk there together.'

I simply cannot get over the synchronicity and the turn of events. Things like this just don't happen to me. Then again this year, they *have* been happening. If I hadn't walked into the cafe, if I hadn't met Deborah, if she hadn't bumped into Sally… it's like I've started a cosmic chain reaction. All I know is that I set out from

London with no plan and, in the spirit of adventure, decided to see what would happen if I let the land guide me on my journey to Iona. Well, this is what has happened. At the last minute, after trying really hard and then giving up, mysteriously favourable circumstances have aligned to help me get to a sacred place; a sacred place I only heard about in an unexpected email on the train ride up.

To me it feels as though the land and I are engaged in a kind of pas-de-deux – neither of us quite speaking the other's native language, but trying to find a common ground. It's inexplicable by rational standards, but it has been happening from the moment I declared my intention to enter into the spirit of my search, for the magical in the land. *This* is what I've hoped for all along. This is a glimmer of the kind of relationship that ancient peoples, indigenous peoples, the true shamans of the world have with the land. A two-way relationship. *It works. It's real.* I have never felt more alive; it is as if some great, gaping maw within me has been filled. I am going to experience a land temple, a myth about to be made real.

'Sometimes things conspire and the timing is right,' says Sally. Like Deborah, she is happy I'm happy, but she's not in the tiniest bit surprised. But then neither of the women have spent four days doggedly searching for a temple in the land.

After the build-up of the previous few days and all the discouraging noises people have made, it is almost comically easy to get to our destination. We walk behind the nunnery to the end of the road. It peters out to a path and we follow it. We come to a gate, go through it and walk on a narrow dirt track that cuts right across the island like an arrow. We pass farm fields filled with sheep, and carry on till we meet the road to the Machair. I'm on familiar territory now. But we don't walk onto the beach.

'That's Cuiberg Farm over there,' says Sally, pointing her walking pole towards a homestead surrounded by a long fence. By the time we reach it, I've seen the gate with a latch. Beyond the farm is another gate that opens easily and a cattle grid, leading to a valley. We are escorted into it by a herd of curious, gingery, long-haired Highland cattle. We pause on the threshold, for this is it: the Glen of the Temple.

The first thing I notice is the deep reservoir of silence, as physically tangible as a bird or a hill or a tree. There are no human footprints on the muddy track. The valley conveys the impression of having rarely witnessed human life. There is no tension here, no grievance or edge or chill. Nothing has jostled this valley. It is entirely at peace with itself. There is a clarity here that I have only experienced before in isolated valleys in Nepal and Pakistan. I remember what Sverre had said, about the pure energy fields in these pockets of land across the island, and it all rings true. A sparrowhawk reels overhead, as if enacting some primal ritual that it undertakes when a human actually does enter the glen.

The ground in this thin place is muddy and boggy and the cows have now fallen behind. To one side I can see the remnants of a long, low stone wall. Sally points to a rocky hill. 'Signal Hill,' she says, whispering. We are all whispering now. The ground becomes drier as we climb up and up onto the crest of the hill. At the top is a spongy bed of purple heather. From here, looking down, the valley is a green strip of runway, bound by grassy outcrops, with the farmhouse just beyond it. Otherwise it leads quietly and unremarkably to the sea, now a choppy grey-blue. The clouds are back, hanging low in the sky, but right above us is a small patch of blue.

The magic, the inspired guidance that has led me here, feels like a breakthrough for me, confirmation that communion with the land is possible if we are willing to let go of our scepticism and our limited perceptions about the nature of nature. It's all so simple: if we let ourselves engage with the Other in the land, the Other will gladly respond. Believing is seeing, as they say.

I am in a state of euphoria, transformed. The glumness that has been hovering over me has now been washed away as if it never existed. I've found new bearings, a way to connect with a heaven on earth, or maybe Gaia as she is when not encumbered by our limited imagination. I may be physically a little cold and damp, but for once, emotionally, spiritually, I'm exactly where I want to be.

'The energy field of Signal Hill and Glen Temple corresponds to the frequency of the throat chakra,' says Sally, breaking into my thoughts. In Eastern traditions this is the whirling energy centre connected with communication. Unblock it and the words, the truth will flow more easily. Could lying down on the grass on Signal Hill looking down over the Glen of the Temple help me find my voice? It's worth a try.

Anthony Thorley had said: 'These landscape temples, with their potent energy fields, have an effect on our vitality.' The thought makes me smile: I had mentioned to him that I have writer's block. Is this why he sent me here? 'If the land hadn't wanted you to, you would never have found the Glen of the Temple,' he will tell me later. It seems I have passed the initiation. Maybe I'll be able to write again, and speak aloud now without quivering.

We lie in silence, all three of us, and Sally and Deborah quite unselfconsciously break into song. Are they singing to the land? Singing as one might in a temple? I let the sounds wash over me before we slowly rise and retrace our steps.

# When You Can See Neither Wood Nor Trees

I'm sitting in Karis Petty's living room. The autumn sunlight is streaming in from her Sussex garden and the anthropologist is telling me a story about the time she'd taken a psychic medium for a woodland walk. 'She had congenitally impaired vision,' she says. It had started out like any other of the walks Karis did as a sighted guide, only this time, the woman (she calls her Amanda though that isn't her real name) began to be aware of unseen energies in the landscape.

'Amanda's skill was in communicating with the dead,' she explains. Being a medium, of course, this was not so surprising. When they reached the woods Amanda pointed to a fallen tree that she could not see. Guided by Karis so that she wouldn't stumble, Amanda explored the tree, feeling her way round it. She 'saw', she said, a dead man by it. And she could hear a wolf howling. Odder still, Karis could hear the wolf howling too. Of course there are no wolves in Sussex. At least not in this space and time.

The medium was thrilled that Karis could hear the wolf because it was proof that she wasn't making it up. But they met dog walkers on their walk shortly after who had heard nothing. It was a mystery. 'Could it be that we are connecting to a wolf that is roaming these woods, right now, but in another dimension?' This is what the two women – one of them the academic before me – asked themselves. In one of her university papers, Karis used academic language to talk about such things: 'an environment sedimented with feeling tones of past activities that are sensed as "energies" with which one can interact and alter.'

Karis had also been to a conference in Finland dedicated entirely to exploring the uncanny in the landscape. 'Do landscapes have inherent qualities that we experience as uncanny or is this uncanniness a product of our perception of the landscape?' This was the sort of question the academic delegates lobbed at each other. Who cares if no one knew the definitive answers? All the joy was in the asking.

I wonder if Amanda would have been able to hear the wolf if her sight wasn't impaired. Would Karis have been able to hear the wolf if she had not been with Amanda? How come the dog walkers, and indeed the dogs, had sensed nothing? Of course Karis doesn't have all the answers, but one explanation Amanda gave her was that she sensed that Karis also had mediumship abilities. And aside from the psychic medium, some of the people Karis has walked with have experienced an eeriness, a profound sense of being watched in the woods too, she tells me. I'm all ears.

I'd found Karis via a blog she wrote for the Woodland Trust. On it she said she'd also guided unsighted people who'd experienced trees as 'wise, anthropomorphic beings' and who described the trees as having a presence. Those hikers were already attuned to the 'brittle rattle of beech in autumn, with leaves refusing to fall' and 'the creaking boughs of the ancient cedar revealing its height', so this is something more than that.

Karis shared too about the mystery of echolocation, listening to the way that sound echoes. An echo, it seems, can tell you a lot about the nature of a woodland. And then there is the rain. 'Rain gives a voice to everything it touches,' she wrote. 'When it rains, the woodland sings. Each leaf sounds a note in this symphony, and the contours of the body of trees reveals itself.'

I'd learned that Karis – Dr Karis Jade Petty – specialises in 'sensory perception and the experiences of environments'. As part of her doctoral research she had explored how walkers who are blind or who have impaired vision experience the natural landscapes they walk in. 'There is so much more to the woodlands than what is seen,' she wrote on the blog, which turned out to be a very good read.

Karis is intimately familiar with the challenges faced by a walker with impaired vision – and there are a lot, not surprisingly: the strain of having to be constantly alert to non-visual clues; a sense of compromised independence and connection; the stress of unknown and uneven terrain to be navigated and the branches, hedges and stones to trip up on, as well as the ditches, nettles, livestock, leaves and puddles; disorienting weather conditions; snow that masks footsteps. One of Karis' walking companions told her that the landscape just felt like a 'load of space'.

Is that how it would feel for me if I couldn't see? Would I have felt the same kind of rapture walking in the fields near the Tor? Or along the Lindisfarne coast? Would I have felt the wind whipping my face and the birdsong more keenly? Savoured the sea air and the smell of seaweed with more care, or wrinkled my nose at the pungent scent of fish and decay? Would I have been able to navigate the labyrinth in Cornwall, never mind listen within for answers to my questions? Or conversely, would I have been pulled into a more meditative space, without visual distraction? Would I have derived more pleasure from my dunking in St Helen's Spring, or had a greater sense of the presence of its guardian spirit? Without sight, would my other senses become like a superpower, heightened to compensate for the lack of vision? Would I 'see' an imaginary landscape in my mind's eye? How would it differ from the physical, tangible one I was walking through? What is a landscape if you can't see it? And is it easier to experience a connection with the spirit or spirits of the land – if you believe in such a thing – if you're visually impaired? I mean, that was the case for Amanda, but then she was a psychic medium.

For a moment, I wonder about all the books written on nature and landscape, all that waxing lyrical – how must it feel to read them if you have never seen hills or mountains or streams or rivers or flowers? I don't have any friends who are blind or visually impaired. If I did I'd talk to them or walk with them. But I shy away from asking Karis if I can speak to one of the unsighted or partially-sighted walkers she volunteers with. I lack gravitas, the rigour of an academic. Anyway, even if they were willing to talk to me I am afraid they'll think I'm trying to romanticise what must be a difficult and challenging situation. And partly I'm afraid Karis herself will say 'no', and think I'm making too many demands or using people for my own purposes.

I decide the only way to find answers to my questions is to go for a hike and shut my eyes. Either that or I'll wear a blindfold. In my head I have this idea – romantic and naive – that I will very quickly develop some kind of sixth sense, like a character in a Marvel comic. I'll need a guide, though. When Karis finally offers to be mine for a walk in the South Downs, I am thrilled. It is a beautiful if underrated landscape – great swathes of countryside where the rollercoaster walks take you over hills and chalk cliffs, where you can lose yourself in bluebell woods and get your lungful of sea air. I am thrilled too because she is working hard on a teaching fellowship in a university and it has taken me months of emails and nudges and patience to pin her down. I feel a little guilty about taking up her precious Saturday and I hope the weather will oblige. She will simply ask me to shut my eyes, Karis says, I'll pull on my hiking boots and away we'll go. I can chat with her and ask her questions before and after our walk.

Polegate in East Sussex on a crisp late-autumnal Saturday is gorgeously bright and perishingly cold. There has been frost on the ground and it's the start of a cold spell, but the sun acts as a psychological buffer against the sharp bite of the wind. I almost feel like I'm back in Montreal when the seasons are turning. The time of year when the leaves are flaming licks of scarlet, gold and yellow, and orange, when you can smell the smoky, wood scent of bonfires, when the school year kicks into a more serious gear and you wonder, half-dreading it, when the first snowfall will arrive, bringing chaos and glee and stoicism all at once.

Karis meets me at the station and thankfully she isn't the stiff, intimidating academic I'd feared she'd be – quite the opposite. She is dark-haired, sylph-like and gentle-natured. In the flesh she looks more like a student than the teaching fellow that she is. A young self-possessed woman with the weight of academia behind her.

'We're going to Folkington,' she says, ushering me into her car for the short drive. The quiet, backwater village where Karis lives is in a valley in the South Downs, surrounded by those lovely hills and panoramic views that are effortlessly lovable. *Views*. How could I possibly enjoy this drive in the same way if I could not see the views?

'What does looking for the "magical" in the land mean to you?' Karis asks me suddenly. It puts me on the spot and it is such a big question. I worry that my answer will sound dumbed-down, not clever or erudite enough. To me magical means… magical. But she doesn't ask me as if this is a test or as if she wants to catch me out – I think she genuinely wants to know what I think. So I pause to collect my thoughts and then give it my best shot:

'It means I'm looking for what I cannot see but which my heart tells me is there and alive and sentient. I want to connect with the spirit of the land. I want to feel heard, cared for, led. I want signs, synchronicity, the whole deal! I want to do a kind of dance with the landscape that both I and the mystery that lingers in the landscape understand on some level, even if no one else does. I crave that intimacy.'

She nods. I think she gets me. Is she going to ask me what 'wild' means, I wonder? I've been asked that before.

'In a natural state' is what springs to mind. But what does 'wild' mean to me? Going feral? Living in tune with the rhythms of nature? Following your desires? Liberating yourself from fear and the expectations of others? If so, this year I'm living wilder than I ever have. And what does 'nature' mean? Everything that's not man-made? The natural world, including humans?

The year before, I'd had the opportunity to meet the musician Nitin Sawhney. Aside from his own music he writes scores for films, including wildlife documentaries. I'd asked him what nature means to him. Quick as a flash he'd replied: 'Nature is everything. This whole room,' pointing to the sterile hotel lounge we were in, 'is an organisation of nature. It's still nature. Nature can be cruel. And the way we manipulate nature can be selfish. But it's still nature. All the bad is nature as well as the good. I think sometimes there's a Disneyfied perception of what nature is' – here I winced a bit – 'I think nature is visceral and at times alarming and dangerous.' He wasn't a tree-hugger, that was clear, and he admitted as much. 'There's a quote in *Zen and the Art of Motorcycle Maintenance* where the author says Buddha is not to be found only in the petals of a flower, but also in the circuits of a computer, which I think is absolutely true.' I admired him for his honesty.

At the end of a steep lane is Karis's beautiful, open-plan home, nestled peacefully amid trees. It is bigger than I'd imagined it would be, the kind of rambling home that successful city folks might escape to on the weekend. A quick perusal of the books crammed onto her shelves – books on nature, eco-spirituality, the environment, Buddhism and mysticism – tell me that

I am on safe ground and can relax with her. She beckons me to the sofa, before pouring lavender tea from a giant vintage china teapot. It's a nice touch. We lean back against the cushions as her cat prowls behind us, miaowing loudly.

'My mother is a medium,' says Karis casually. I'm fascinated. It explains why she is so open-minded: when your own mother speaks to the dead… Maybe this is why she's interested in exploring, as she puts it, 'the uncanny in the landscape'. I understand this a little. My own maternal grandfather, who died before I was born, was a healer and used to perform exorcisms, my mother once told me. And I had an aunt who was a clairvoyant. 'Maybe that's why you're interested in experiencing a more spiritual dimension to life,' says the academic. Maybe. I'd never thought of it like that before.

Before long it's time for our walk and so we set off back down the lane. Will we be walking on the Downs, I wonder? Will Karis take me to a sacred spot? How will I know if I have no visual clues – no stones, no cairns, no signs? And what makes a landscape sacred anyway? Is it a place where rituals to connect people and spirits once took place and still do? A place of peace and inspiration, where there is a special energy?

Sacred is another of those words that slips through my fingers. In the past, I've experienced the sacred when stillness and a beautiful landscape come together and I no longer know where I end and it begins. Or in the musky incense of temples and churches, with their shrines to the divine. I've experienced the sacred in the prayers and blessings that people have offered in these spaces and which are tangible in the ether. I've experienced it on the uninhabited islands in Sweden where I camped tentless, and in a night sky ablaze with stars over Queensland, in the dancing Northern Lights in the Arctic, in the astonishing vastness of the wilderness I flew over and into in Namibia – in countless landscapes, in fact.

But I also experienced the sacred as a teenager when I was in a hospital at my father's side after he'd had a heart attack. The love between us – deeper and bigger than I'd ever felt before – had felt sacred, as had both his and my sister's sudden deaths, even with the shock and trauma. For death too is a sacred passage.

Yet, when it comes to a landscape, sight seems to me integral to a feeling of sacredness, whether you are high on a mountaintop in Nepal or shivering in a valley in Sussex. Then again, I have spent my life with my vision intact. Maybe – I hope – I am about to be proven wrong.

On our walk, there is to be a sort of protocol. 'We're not going to talk,' says Karis. Silence is going to be crucial to my ability to listen, to feel, to dwell in this unseen landscape. I often crave silence and so I tell Karis I have no problem with this. She nods: 'Most of the unsighted people I've taken on walks have said they wished their sighted guide would stop talking so much!'

There's no blindfold; I'll simply be trusted to close my eyes and keep them shut. With one hand, I am to lightly grasp a spot above Karis' elbow as we walk. 'I'll be scanning the area for you, so you don't need to worry about tripping and falling,' she says reassuringly. She'll break our silence to tell me if we are approaching a bump or a fallen branch or anything else that might trip me up or stop me in my tracks. Then, with my free hand I can follow her arm down to her hand, so I can get a sense of whatever lies in front of me.

My eyes are still open as we stroll through the village. I'm hungrily taking note of everything I see: a few houses, a pavement, a church. Will we be sticking to the village paths? Or will she take me into the hills?

'OK, we start here,' says Karis, stopping suddenly. We have yet to come to a crossroads or a stile or anything that feels to me like it might be a significant marker in the landscape, something that might give me a clue as to where we're going.

'Can I tie my shoelaces?' I ask. I'm nervous now, stalling. Once I've undone and redone my boots, I close my eyes, the terror of cheating keeping them welded superglue-like. We set off at a pace.

'Am I going too fast?' asks Karis. I nod. I'm a slow walker with my eyes open, never mind shut. I'm a clumsy one too, so I'm glad when we slow to a crawl. The warmth of the sunlight on my eyelids surprises me in the way that looking at the sun never does. I'm acutely aware of the ground, solid and rising to meet my feet. I can feel the earth, generous and supportive. I feel the invigorating crispness of the air, and it is a lively, electrifying companion.

We come to what I guess is a gate. I pause while Karis opens it and shuffles round the other side with me, onto what must be a footpath. And yet I can hear cars on a busy highway. They sound very close. Why is Karis taking me for a walk near a highway? I didn't come here to hear the roar of traffic. My feet are treading on rough but level ground, and my mind is whirring in a way that it wouldn't if I could see. Walking when you're unsighted demands a lot of thought.

I hear a glimmer of birdsong up high. *Ca-caw*. A crow! And then the flapping of wings, somewhere above my right ear, like a missile swooping past. The sound animates my surroundings, which right now feel like a narrow footpath and open space and warming sunlight, and those annoying cars. I have no perception of hills or trees. I'm not too sure where 'out there' begins and 'in my head' ends. I feel that I am hovering in darkness, on the brink of falling backwards into myself. I'm not sure if this is a scary thing, a good thing, or just a little weird. Karis' arm is reassuring, and I can feel a gentleness emanating from her and wrapping itself around me in a soft cocoon. She'll make a good mother one day, I think to myself from nowhere.

Every so often, she remarks on the terrain: 'You're coming to an uneven bit, a kind of furrow.' Or: 'There's a puddle here, about a footstep in length, so you could jump over it.' Or: 'We're heading downhill now.' As we walk, I try to shift to a lower centre of gravity, like a dancer might. When I do, it's like I have a second brain down there, guiding me. There is no palette of greens, no hills, rolling or otherwise, no blue skies, cloudless or otherwise – the land of the unsighted is a cliché-free zone – but instead, a delicate dance of poise and balance. Everything happens in relation to my inner landscape. Even a breeze lightly whipping my face feels like a personal greeting. A dog bounds up from nowhere and its tail is hitting my leg eagerly and affectionately. I love the warm, friendly energy of this dog.

I still imagine I am inching along in an open field, but then the light disappears and I wonder if we have entered a wood. It feels big and mysterious, the ground underfoot a little sticky. 'There's a branch across our path,' says Karis. I reach out for it – it feels as dangerous as a snake – before it can ricochet across my face. Then I trace it with my hand before I abruptly stop. Why have I stopped? 'Good instinct,' says my sighted guide. 'You stopped at a barbed-wire fence.' Oh, not a wood, then.

Now I come face to face with something hard and cold. A concrete wall? But what is a concrete wall doing here? This is a very strange landscape, I think. Inching my way round it and crouching down to the foot of it, my hand glides across cool, moist moss and then my fingers grapple gnarly roots. Ah, not a wall at all but a tree. A tree with a wide girth. But what kind of tree? I feel the bark. Very hard. Few fissures. *What kind of tree is it?* It maddens me that I can't identify it. I can't see or touch its upper branches or its canopy, if there is one. Is this a tree with its branches open wide and wild like Kali? Or a more upright tree, say, like a poplar? Do poplars even grow here on the Downs? I let go of Karis' arm and creep round the tree some more and feel another tree brushing against my back. Solid as the earth, supportive, here for me. I can nestle against one trunk and touch the other. Karis meanwhile is silent. Is she watching me? Is she impatient with me, willing me to hurry up? I can't see the expression on her face, so I don't know.

I turn my mind back to the trees. They are a friendly presence. I stand still between the two, all interest in deciphering the landscape 'out there' forgotten. I am being pulled into a world that is made up of me and this slow quiet being. It's quite intimate and subtle, and to enter into it fully I would need more hours than I have, but even in this short time I feel the tree respond to my tentative explorations. I have the sense that it is holding its breath, not wanting to disturb me. In its tree-slow way it may even be enjoying the attention I am giving it. This tree is receiving me tenderly.

There is no wolf howling, though, no sense of a non-human, non-tree presence or spirit. I stand still for a long while. It's quite pleasant, standing here, deep within myself and this darkness. If I were here alone, I think I would stay and stay. But then if I were alone, I'd never find my way back. The sobering thought pulls me out of myself.

'Hello?' I call out to Karis. 'I'm ready to go.'

I feel her hand guiding mine to her shoulder, and then she leads me along some kind of zigzagging ditch or furrow for a while until we emerge into the sunlight again. Never has the sun felt so good, so much like tenderness, on my eyelids and on my face. 'Welcome to my abode,' I hear the sun, a silent whisper in my head. The noise of the cars has long since died down.

'We're passing a sacred spot in the landscape,' says Karis. 'People gather here during Beltane.' It's the spring festival, a day for lighting fires and dancing round maypoles, she tells me. Her words have me imagining I'm walking on a path cut into a hillside, and that perched in the hills to one side of me is a standing stone, or maybe a cairn. I don't feel any sense of the sacred, for I am too busy convincing myself of the shape of the stone that I can vaguely see in my mind's eye, with my superior powers, my inner sight. I'm almost certain it is there. And then we stop abruptly.

'We're here. You can open your eyes now,' she says.

I do as I'm told. There's a blast of light on my retina, a deep green field tumbling like a wave down the valley, clear blue skies painted with a wide brushstroke. I am astonished. With my eyes shut I'd been walking through a smaller, more shut-in space. I'd not realised how deeply I'd sunk into the recesses of my mind. The contrast of this panorama is shocking. Though I *am* walking on a footpath cut into a hillside. I'd got that bit right.

'Turn around,' says Karis. I do, slowly.

'Oh my God,' I say, gasping in astonishment. Behind me is a chalk figure etched into the land, the size of forty men. She has brought me to the Long Man of Wilmington. I wasn't expecting this. The Long Man, guardian of the South Downs, is a figure that baffles historians and archaeologists and causes unsuspecting walkers to stop and marvel. Is it a prehistoric fertility symbol or a more recent folly? I stand and stare for a long moment. I've never seen it up close.

I am so thrilled to be here and so relieved that I no longer have to rely on my imagination, my muted inner radar, the on-off feeling of connection. But then again, if your inner vision is all you have to go on, you are creating an entirely unique landscape as you walk, one conjured by your psyche and your senses. It's a wholly different landscape to the one you see with your eyes. So which one is more real? Is a landscape a subjective thing that only exists in the eye of the beholder? I have no idea, but I have enjoyed asking the question.

I take a flask out of my rucksack and pour hot ginger and lemon tea for us both. I hand Karis a cupful and she wanders off and gives me some space to scribble notes – not something I've been able to do for the last hour and a half. It has been a strange, disorientating experience, but I sense that my sightless impressions will be lasting ones. And this proves to be true – at least, months later when I write this it is still true.

I can see Karis walking towards me. She must be freezing in her sweater. I'm freezing in my down jacket. 'Do you want to keep your eyes open on the way back so you can see where we've been?' she asks.

'Yes please!' I say. I can't bear the thought of shutting my eyes now that I have opened them. The walk back is enlightening. That funny little furrow? It's just a footpath in the hillside. The deep, dark, eerie wood of my imagination? A tame cluster of trees off the footpath. Here is that branch that I'd felt across our path. It looks so ordinary now, nothing like the looming, frightening, python-like gatekeeper of my mind's eye. And the tree or rather two trees I'd nestled between. To my great shame, I still can't name them. There are white splodges on the wide, smooth trunks, which are grey as an elephant's, so I think it might be a beech. It's impersonal now, 'just' a tree, a little distant, its spirit under lock and key. I cast my eyes towards the ground and pick up its golden, oval fallen leaves, like I'm an arboreal Sherlock Holmes. I can't be certain. We walk on and soon we come to a great field of grazing sheep. 'Did we come past here before?' I ask.

'Yes, they were silently observing you,' she says. And yet I'd not had any sense of their nearness, I'd not heard them or scented them – can you smell a sheep in a field? I sniff the air. Maybe a faint wooliness, a grassiness. Not really. I had no idea that we were even near an open field. We cross the field and another stile and we are back in Folkington. I have no idea why I thought we were near a highway. We are nowhere near a road. How strange. The sound must have travelled up from the valley. And yet I can't hear any cars now. Not a single one.

Back in the house, Karis peels and chops a squash and makes me home-made soup that we eat with hunks of bread around a cosy fire. I am glad I can feel its warmth and I'm mesmerised by the flames – flames I can see. I had walked and hoped to experience the landscape in a mystical way, the way the medium had. But I'm not a medium. My encounter with the tree, though: there was a true, wordless connection there. All in all, a mixed experience.

'But you tried, you tried to perceive the world in a different way,' says Karis soothingly.

That night, back home, warm and curled up in bed, I google 'blind hiker' and come across a man named Trevor Thomas, also known as 'Zero Zero'. I read that he was the first unsighted person to hike solo, unassisted, along the US Appalachian Trail. All 2,175 miles of it. Mind-blowing.

I put my laptop away and think about how I ended up standing beneath a figure that evoked mystery in the landscape. Was that not a magical thing to have happened? Karis may have taken her cue from me in leading me there, knowing I was after an experience less ordinary, but why this feature and not another? Maybe the wild Other had been guiding our walk the whole time.

Suddenly it matters less to me that there are those who will think I am mad for harbouring such thoughts. It matters less, I realise, because this year I have been having the time of my life.

# In Search of Ash Dome and Maidens of Mud and Oak

It has rained for weeks – bitter, wet and torrential – and now finally the sun has returned, full beam as if to make up for its long absence. Officially it is spring but, aside from the flute-like calls of the birds and a few buds and snowdrops, you'd never know it. My own weather system is currently set to 'gloom'. I've barely left London in the winter months – though I have, most days, walked in the woods. I've not let up. But I am also dealing with my GP's thinly disguised portent of doom. Hearts are weak in my family, but I hope to buck the trend. I have nature on my side, my beloved.

Still, I'm sick of hibernating and I need something to blast the fear into oblivion. I hope and pray Ash Dome, the artist David Nash's tree sculpture, might be it. 'A circle of life made from life itself,' intones the handsome presenter on a BBC programme about land art, walking round the trees, caressing them. Ash Dome, he tells us, is 'a living artwork, a closely guarded secret, in a woodland somewhere in Wales, its location exceedingly difficult to find'. Closely guarded secret. With those words, he's unwittingly thrown down the gauntlet. If this graceful, winding form of 22 ash trees is a closely guarded secret then it is one I am going to find my way to.

I have a soft spot for land art – that is, art made from things in nature and, once finished, exposed to the elements. I love the way these strange, striking creations often appear so unexpectedly in the landscape. They are created from the land but then take on a life and spirit of their own.

I once came upon a medicine wheel crafted from rocks in a river valley in a remote part of eastern Iceland. The artist, I discovered, was a First Nations elder from Canada. The wheel, he told me, when I tracked him down, was created to honour and heal the earth. 'It has great power,' he said. He'd laid the unobtrusive ceremonial structures down all over Iceland.

Then there was the astonishing work of art in the desert in northern Namibia that I discovered. I was on a work assignment and had taken a tiny plane up there, close to the Angolan border. The landscape was a strange, hallucinatory pinky-orange, bound by stark mountains. It was Mars on Earth. One day when I was out with a guide, I'd come upon the strange, flat, biscuit-coloured circles made of rocks. They imbued the already extraordinary landscape with even greater mystery. The sculpture, called 'Sacred Fire', was created by an Australian artist by the name of Andrew Rogers, together with the local Himba people. It was part of a global land-art project called 'Rhythms of Life', apparently the largest contemporary land-art project in existence: 51 stone structures in 16 countries across seven continents.

Up in Australia's Queensland, I saw the Quinkan rock art, named for the ancestral spirits said to be hiding in the rocks. The mysterious ochre drawings of stick figures and animals are scattered across sandstone boulders and caves in a remote tree-filled valley. I don't know if they depict hunting stories or a medicine man's journey into another world, but they held me rapt.

Even as a child I had a fascination with land art – some of our greatest mysteries revolve around them: the Nazca desert lines in Peru, for starters. Who created them? What did the lines mean? I'd spend hours wondering. Was Stonehenge some sort of prehistoric art? I'd read that ancient carvings invisible to the naked eye had been etched into the stones, which would make Stonehenge an art gallery. I liked ephemeral land art too, like Andy Goldsworthy's creations formed of flowers and twigs and leaves, stones and beach clay.

And now I am on the trail of the love-child of a human and a tree. Ash Dome was planted 40 years ago, in a private woodland called Cae'n-y-Coed. The word translates to 'field in the woods' in Welsh. I do some research, put feelers out, and eventually my persistence pays off. The ping of an email brings good news from David Nash's studio director Nia: '"Ash Dome" is in a private woodland and not open to the public, but David is amenable to you visiting.'

* * *

In Blaenau Ffestiniog, my guesthouse is opposite the train station. It was built of Welsh stone and slate in 1870 by the local doctor. A laminated menu in the cafe tells me he was a poet, musician and fisherman. From where I am sitting, I can just about glimpse the mountains, stripped bare from the mining – an industry that once filled the streets and houses with workers and their families. Even though Blaenau is geographically smack-bang in the middle of Snowdonia National Park, officially it isn't a part of the park at all. Not enough of a beauty spot, it would seem, thanks to those bare mountains, which look even more oppressive under the gloomy skies – sadly the sun has not followed me from London.

Tomorrow Nia will tell me that the townsfolk wear their outsider status with a certain pride. This evening, though, I am on my own. After I'm shown my room, I dump my things and walk down the empty high street for a recce. The cafes and the few shops on the main drag are closed, and the village feels cold and lonely. The handful of pale-faced teenagers I pass on the high street have a pinched, expressionless look about them, the very essence of 'nothing to do and nowhere to go'. I am a bit anxious about what might come out of their mouths when they see me.

I turn down the first street I come to, and end up in what was once an Edwardian police station but is now some kind of arts hub – not what I expected to stumble upon here. Up a staircase, I find myself in a quiet room with vaulted ceilings and a sort of shrine to Bob Marley at one end. There's a roaring fire and jazz oozing out of the speakers. What luck. What a relief too – I'd been too nervous to enter the pubs I'd passed, what with their aura of white tribalism, a force field that keeps me at bay.

After a dinner of chickpea curry served by a sweet, smiley waiter, I walk back to the guesthouse and consider my plan for the following day. I still don't know where Ash Dome is. But I will walk over to David's studio in the morning. From there, because David is not well and not in Wales at all, Nia will take me to the woods. I can't help but feel a little like a character in a thriller who is to be escorted, blindfolded to a secret lair. I feel a thrill of anticipation even though I'm disappointed that I won't get to meet the artist himself, and I decide to spend the rest of the evening going over what I have learned so far.

★ ★ ★

'I think what lies behind everything, particularly in nature but also man-made and creature-made are spirit forces. We have an instinctive grasp on that, but everything is a threshold into something beyond, into a parallel universe. It's very connected.' I'm curled up under the duvet, reflecting with some satisfaction on the sculptor's words.

When I emailed David, he'd shied away from talking to me about his love of trees in spiritual terms – 'too loaded', he'd said – but then I found this snippet online and sent it to him. 'Oh good, that's a good one. Keep on digging, there's lots out there,' he'd replied. I could almost hear him chuckle.

I reread the other scraps I've amassed: David grew up in Epsom and was interested in making art from wood from the time he was a child. He had a profound respect for the wisdom of trees and later came to call them his guide and teacher. His father had bought a home in North Wales and all of David's school holidays were spent there. He went to art school in London, and was eventually gifted a woodland by his father, which would later become the site of Ash Dome.

'In North Wales there was a strong sense of land and weather. We were there winter, spring and summer. There was a strong sense of the seasons,' the artist said when I asked him what it was about that landscape that had had such an impact on him that he'd chosen to return and live here. He'd bought a chapel in Blaenau Ffestiniog and turned it into his studio, the one where I am meeting Nia tomorrow.

Ash Dome is a wonder – that comes across even in the pictures I've seen. You could call it a love story between a man and his trees. An interspecies collaboration. Human life meeting non-human life and creating a new, extraordinary blended life. It was conceived in the 1970s to mature in the 21st century: an act of faith. David Nash wanted an answer to a question: how to make an outdoor sculpture that is genuinely of its place and can sustain its freshness over time? It was, he said, 'a gesture in a very dismal and depressed global time, of hope… that the human race would survive into the 21st century.' I consider that for a moment. We have, just.

This living artwork is the result of a vision, a million tiny labours, patience, craft, and a coaxing, an engaging with the qualities of the tree, the desire of one man to immerse himself in the land and bear witness to the changing seasons year in and year out. And to learn to speak 'wood'. Not surprisingly those who have visited speak of it feeling like a sanctuary, a place of rest and reflection.

What I haven't yet a clue about, so focused am I on visiting the secret woods and the ring of trees, is that entering David Nash's studio is like walking straight through the looking glass.

It is not often that you enter a parallel universe, but rocking up the next morning at Capel Rhiw after a short walk, I am sure that I have. The former Methodist chapel in which his studio is housed is striking enough, but when his studio director opens the door I can barely believe my eyes: imagine entering the Sistine Chapel or Tutankhamun's tomb and that is how I feel: a kind of awe. I am staring into a hall filled with Mad Hatter-esque creations.

'Welcome,' says Nia, who has a kindly look about her that makes me want to give her a hug. She ushers me through a portal made from what looks like a giant overlapping wishbone. 'One half is carved from the wood of a "frêne" – the French word for an ash tree – the other from a "chêne", an oak tree,' she says. 'David calls them "The Ubus" after Ubu Roi, a comic character from a French play.'

I nod and smile, my head reeling – I'd read *Ubu Roi* when I was a student in France, which makes the reference all the more surreal. Here is a forest of inspiration, hewn from naturally fallen wood: tall thrones and totems, quirky giant-sized spoons, smooth eggs and strange-looking hives, a larger-than-life tree tongue, what looks like a canoe, a tree-organ, wood clams, boulders with splits in them, recurrent patterns that are weirdly pleasing to the eye, hollows, and waves, and rounds. Playfulness and grandeur and solemnity and the absurd are all here – camouflaging skill and vision, and – whisper it – genius. Sunlight bounces through the chapel windows and creates a rainbow when it strikes a charred sculpture that resembles a thick, fat, upright python. I gasp. Nia watches, delighted by my reaction.

She's clearly a fan of her boss's work and talks lovingly of the sculptures as we weave our way in and out of them. I struggle to remember any of what she is saying, though, for I am dizzy and in danger of losing sight of the trees for the wood. 'Quite something, isn't it?' she says. I nod, speechless.

After tea and biscuits, we make our way to the wood. The wood that David Nash cleared and planted and has loved since the time it was given to him a half-century ago. I am finally going to see Ash Dome. I am sworn to secrecy about the exact location, but we enter the wood through a path lined with trees that beckon invitingly. The sun chooses this precise moment to burst through the clouds and it feels like a benediction, nature shining a spotlight on this fine creation.

'In just a month, the bluebells will be out,' says Nia. We are walking up the path and then suddenly in the clearing there it is: 22 slender ash trees dancing in a circle; 22 trees in communion, against a backdrop of mountains. A ring, a circle, a vortex, a living monument. I suddenly remember the print of Matisse 'La Danse' that I'd hung in an old flat. Ash Dome reminds me of it.

I can feel the trees inviting me to join the dance: a dance in honour of the elements, of life. This impression is down to the kinks in the trees, forged over the years David has spent training them. That and something more. It is as though, through his sustained passion and commitment, the artist has conjured a third being, a spirit, a genie, a heartbeat. Ash Dome is a kind of shaman tree, a liminal space between two worlds: the world of physical nature and the world of man.

'Go and take a closer look,' says Nia. 'I'll stay here and give you some time.'

So I walk right up to the ring of trees. The tops of the ash are swaying in the breeze. They look a little wild and bushy as if they are in need of a haircut. It is early in spring and they are not yet covered in leaves. I walk round and round, weave in and out, lie down in the middle of them and look up at the tree tops creating a star pattern above my head. I feel as though I am levitating.

'How to make a sculpture that did not seem alien, an imposition, a UFO – made somewhere unknown and then landed fully formed.' Lying down in the grass, I recall David's words. To me Ash Dome feels wonderfully of its place but also wonderfully Other. But not at all alien. I wish with all my heart that I could spend a whole afternoon here, in the middle of these trees, with the sun pouring down on me.

Around the sculpture are tall beech trees standing sentinel, protecting it from the wind. 'When the bluebells carpet the woods, the trunks of the ash trees disappear and it looks as though they are floating,' says Nia when I walk back to her.

Had I stumbled upon this living artwork on a hike, knowing nothing of the artist, not knowing it was a 'closely guarded secret', would I be feeling the same intensity of emotion? Yes, I think I would. I would have experienced even greater astonishment.

We walk on and Nia draws my attention to more surreal creations: 'Japanese larch trees,' she says, pointing to trees, dark and curvy as serpents, standing or in clusters on a hillside. There are little hive-like mounds composed of lichen-covered logs, and the *pièce de résistance*: 'Seven by Seven', a zen-like arrangement of the whitest, most papery-barked of Himalayan birches, each planted seven feet apart in seven rows of seven, on a slope. Each of the trees is 49 feet high. It's an army of birch recruits, a kind of surreal white cube, a nature-loving mathematician's dream, all clean, geometrical lines, carefully maintained. To David, it's simply a 'planting project'. Wherever you stand, you have a clear line of sight through the trees, and I can see the Moelwyn mountains rising up in the blue sky.

I could spend all day here, I think, weaving in and out of the birches, racing between this sculpture and Ash Dome. I long to secretly sneak back, but in truth I am not sure I would even remember how we got here.

After lunch in a lakeside cafe popular with builders and locals, Nia takes me for a drive along the River Dwyryd and shows me the journey of David's free-ranging 'Wooden Boulder', a sphere carved from oak that took on a life of its own, popping up and disappearing randomly at spots along the river as it was swept along by the tide. It captured the imagination of locals and even people from further afield. 'One woman flew out here from Japan and just sat with it,' says Nia. The last time it was seen was in 2013. Has it become stuck fast in mud under the river? Or is it now somewhere in the sea? No one can say for sure and the boulder's not talking. The seeking of it when it disappeared became a kind of performance art in itself.

Is it the vulnerability and impermanence that makes this type of living art so beguiling, so easy to engage with, because we recognise those qualities in ourselves?

The trees could be swept away by a storm at any time, just as we might.

When I finally walk back to my guesthouse in the rain, clutching a small twig from Ash Dome, signed by David Nash himself, I am exhausted and exhilarated beyond belief.

Of course, the next day is destined to be a crashing anti-climax. In the morning, I wake to great rivulets of rain snaking down my skylight. I'd planned to go for a hike in the hills but this is not walking weather. I can take the train and return home – the sloth-like, rain-loathing part of me likes this option – or I can carry on through the Conwy Valley to Llanberis. I've had a last-minute invitation to stay at an eco-retreat called Cae Mabon overlooking Lake Padarn. I reread the message from Eric Maddern, the founder, on my phone: 'If you want to experience another kind of art in the landscape I think you should come to Cae Mabon. Different to David Nash but beautiful and inspiring nonetheless.'

He tells me a work party is building a land sculpture, a sort of earth goddess, intended to foster a deeper connection with nature. I like the idea of joining in and helping to play a small part in her creation. I flip a coin and, despite the torrential rain, Llanberis wins.

This also gives me a chance to take Ffestiniog's famous steam train part of the way to Betws-y-Coed. The little locomotive is a tourist attraction. Every waterproof-clad commuter leans in to listen to it sigh and hiss and belch its way into the station, looking for all the world like Thomas the Tank Engine.

There's not much in the way of scenery en route, owing to the pounding rain steaming up the windows, and at Betws-y-Coed there's no let-up. Here I get on an empty bus and it rises up and up through stark, coldly beautiful mountains.

The last time I was in these parts I was on a silent wild-camping trip. I loved the idea of a silent hike, but we'd chosen the hottest day of the year and I was shattered by the time we pitched up by a remote lake. The next day, we'd hiked up to the top of Mount Cnicht under blue skies and I'd felt invincible. I'm not feeling that way today. It's not just the weather; anything would be an anti-climax after yesterday.

Besides, utopic retreats can be plain miserable in the rain – the mud churning up the ground, the lack of heating chilling your bones no matter how many cups of tea you clasp in your frozen hands. And there is often a kind of reverse snobbery at work: rucksacks are in, suitcases on wheels (a marker of your mainstream, capitalist leanings) are out. I look down at my red suitcase with its four wheels – with a bad back, rucksacks are for evermore out for me – and shrug. Too late now.

Cae Mabon, also known as the Field of Divine Youth, sits in an oak forest clearing. Most of the homes are on a steep hillside and are handmade from natural materials – straw bales, turf, cob, stone, cedar logs – it's majorly hobbity, all curvy and rustic and low-impact and self-sufficient. I pass a cabin and notice bright yellow daffodils springing from its turf roof in defiance of the grim skies.

The people here call this place a Welsh Shangri-La. I'm sure it is on a hot, sunny day, but today I just think it's going to be bloody cold and damp at night. There are some good walks, though: you can walk down to the lakeside or you can walk uphill and join the Lake Padarn Country Park path. We're in the heart of Snowdonia, and the views of the mountain and the lake when the clouds clear must be something.

But I've come at a bad time. It's the end of the five-day working party, and the group of volunteers have already gelled into a tight-knit clan. 'Talk to people,' says Eric airily, the minute I arrive. I'm led into an overheated communal room, filled with humid, steamy air, screaming children and people – the working party, I presume – who barely glance my way. I'd turn around and leave if I could. I miss Nia and her friendliness. I miss Ash Dome. I miss the slate mountains and the anonymity of Blaenau Ffestiniog.

On the drive over from Llanberis, a more forthcoming Eric had told me a bit about himself: 'I was born in Australia and went to school in England, but I've travelled all over. I was keen to learn how people in other cultures communed with their gods,' he'd said. 'I went through dark times but I found my way back. Nature was my biggest healer.' Later, he went back to Australia and lived and worked in indigenous communities as a community artist. He was inspired by the Aboriginal 'Dreaming'– creation stories, landscape maps composed of song, and art, ceremony, ritual, mythology, and actual dreaming, all of it barely translatable in the English language – to create this radical, utopian commune in North Wales. He'd long felt the pull to settle here. He'd spoken of Mabon, 'the Great Son of the Great Mother' in Welsh mythology, or Modron, the Great Mother herself.

The references to the Welsh mythic landscape leave me cold, in contrast to the Aboriginal Dreaming, which enchants me even if I can barely grasp the notion. It's not surprising: Eric has lived here for years and he has Welsh grandparents and roots here; I don't. Where do I fit in? I'm happier with a pick 'n' mix of myths from various landscapes: stories from here, stories from there. A bit of this tradition, a bit of that. A sort of bastardised mythic landscape that the rootless might relate to. Maybe I relate to the Aboriginal Dreaming because the indigenous people in Australia have been Othered for so long, and I identify with that. Maybe I just hate the phrase 'mythic landscape'. What is the mythic landscape anyway? The features of land, river, sky that become associated with stories of heroism and courage? I have to look these things up, for they are not native to me. What is so wrong, I ask myself for the umpteenth time, with wanting to connect with the land on your own terms? I am all for operating in a world of spirit, of having a direct experience of the magical, of the Other, as the indigenous people from all cultures have done. But which specific culture am I meant to identify with? If you come from a multicultural background, it isn't an easy question to answer.

At Cae Mabon, Eric's redheaded partner looks me up and down coolly and offers me a cup of tea. 'So you're researching the mythic landscape?' she asks politely. No, I reply tightly, I'm not: 'I'm exploring the magical in the landscape.'

A quest for the magical feels to me inclusive, open to anyone of any culture, any background. It could include the mythic landscape of this land, and all the mythical characters and 'players' and stories if you wanted it to, but it doesn't have to. It could be about connecting with mystery, seeing things with the eyes of wonder, the eyes of a child – no cultural baggage attached. That is much easier to access. If you are an urban kid from a tough inner-city neighbourhood, with little experience of nature, or if you are a refugee from a beautiful, distant land with its own traditions, you could get it. If you are a corporate high-flier in London with a yearning for a life more meaningful, you might not get it but you likely wouldn't be actively turned off by the tribal-sounding 'mythic landscape'. You wouldn't worry if you weren't part of the clan that has all the landscape vernacular and myth at their fingertips and who, at times, exude a kind of self-satisfied superiority about their knowledge.

But maybe these are just my gripes, my projection. Then again, when I told a woman I met recently that I was exploring the magical in the landscape, she point-blank asked me: 'But what does that even mean?' She was stumped. And I was shocked. How could she not get it, not imagine, not conjure in her mind's eye…? So all things are relative and I'm operating in a bubble of my own too.

I've only just arrived but I am fed up of this place and what feels to me like tribal cliquishness and the looks my little red suitcase gets. No one has invited me to help with the earth goddess sculpture and I don't think they want me to either. I excuse myself from the overheated room and decide to go for a walk. I need to walk off my anger. Thankfully the rain has let up a bit. I cross a bridge over a little cascading river and manage to find my way through the forest and onto a footpath, nearly tripping over a wan young man crouched on a mossy stone wall, smoking dolefully.

I walk up and up and up till I find a waymarker – it seems I have joined a public access path that snakes around Lake Padarn Country Park. I perch at the top on a slippery stone and stare moodily out at the lake and the mountains it frames, layers of land-clouds surrounding a grey shimmer. It is beautiful even in the gloom, and at any other time I'd be congratulating myself for finding myself in such a place but I am still too pissed off to really take it in. I stay as long as I can, like a delinquent, runaway child, till the feeling eases and I eventually turn back to the camp, down and down and down.

I notice a sculpture carved into the trunk of a big oak tree by a stream. It is a shapely female form and it looks like she is holding a heart in her hands. Whoever made this is a gifted sculptor. I sit on a bench, listen to the rushing water and gaze admiringly at the sculpture. The woman's face, I notice, is framed with curls. The doleful young man who'd been smoking comes over. 'We call her the Oak Maiden,' he says, nodding. 'She was carved into a lightning-damaged trunk. Pete sculpted her. It took him about a month.' Pete, it turns out, lives at Cae Mabon. The oak tree rises high in the forest, and the two top limbs reach out wide as if adopting a pose you might see in an ancient Egyptian mural.

After a while I wander up a hill and come to a rigged-up shelter. Three volunteers are patting and coaxing the clay in the shape of a supine goddess. They are in a kind of trance, totally absorbed in what they are doing. I stand apart and watch and feel like an outsider. I don't fit in here. 'Who inspired the sculpture?' I ask, immediately regretting it for now I sound like a reporter – I imagine they love journalists here. 'It's the Mud Maid at the Lost Gardens of Heligan,' says one of the women, pausing in her patting.

Heligan, an anagram for 'healing', is a vast country park in Cornwall. I've never been and know little about it, but I wonder about this Mud Maid. I wonder all night long, because after a communal supper I sleep in a perishing-cold yurt with about ten blankets on top of me and still freeze to death. I can hear people laughing and joking in a hot tub behind it till the small hours. If only they'd shut up. I cannot wait to leave.

The next morning I drag my red suitcase up the hillside as the clan look on – not one of the young men hanging about helps. So much for utopia. Eric kindly drives me to the station and we chat some more.

'I think of the maiden in the tree as the Faerie Queen,' he says. 'Pete always calls her "the goddess". Modron is a mother goddess. The tree carving is more a maiden. Perhaps Blodeuwedd, the Flower Maiden, is the closest mythological figure to her.' I nod and understand nothing. He tells me he is writing a memoir and I wish him well with it. We don't really talk about the earth goddess sculpture.

* * *

Two days later, back in the warmth of London, I am astonished to get an invitation, with a picture of the Mud Maid attached. It has come totally out of the blue. 'Would you like to come and spend a night sleeping in the Lost Gardens of Heligan? Under the stars? Would you like to meet the Mud Maid?'

This is a mind-boggling bit of synchronicity even by my own standards. The person who has sent the email is a wilderness guide named Danny, who I met on a fasting retreat under open skies in the Sinai desert a decade ago. It was he who had gone hunting with the people of the Kalahari and experienced their near-supernatural ability to sense the presence of game animals. He is organising a UK version of the Sinai retreat in June and needs people to recce it. Of course, I accept. I can't ignore a sign.

★ ★ ★

Heligan is a kaleidoscope of colour and light and heat – God, the glorious heat – five miles from St Austell. I spent the previous night in a sleepy little village called Pentewan that is too tiny for me to feel inconspicuous but there is a beach close by and the landlady at my B&B bends over backwards for me. Nothing is too much trouble, and I'm touched. This morning I get a lift to Heligan. 'You're in for a treat,' says June, whose partner is involved with the Eden Project, and who Danny has corralled into chauffeuring me.

It's the perfect summer's day – not a cloud in the sky – the start of a three-month heatwave, although we don't know it yet. I immediately fall madly in love with the park with its vast meadows and fields, gardens and beehives and sea views, wildflowers and bluebell-covered woodland. It is truly a wonderland. Astonishingly our group will have the run of it when it closes to the public in the afternoon. And the Mud Maid? She is a sleeping beauty in a 'catsuit' of ivy and a crown of spiky grass atop her head. Cornish artists Sue and Pete Hill, who are brother and sister, created her. Nearby is another of their creations – the Giant's Head, which seems to peep up playfully from the ground. He sports a halo of orange flowers. His piercing-blue eyes are, I learn, mosaics made of broken dinner plates and blue milk-of-magnesia bottles from a rubbish dump.

'The Mud Maid is a Green Woman, a woman of the woods, an earth mother,' says Sue. The woman of the woods is over 20 years old and has aged well. She is made from mud, cement and sand, around a framework of timber and some kind of netting, and her hands and face were coated in yoghurt 'to set the lichens growing'. I can't help but feel that it is through the grace of this mysterious being of the land that I have come here.

That night I bed down in a big meadow with soft, tall grass, and wildflowers. I have the meadow entirely to myself. I spend long hours obsessively tending to a fire, feeding it the kindling and the charcoal I've been given. I fall asleep with the embers glowing, under a velvety sky a-frenzy with stars. The witching hour is upon me. There's a cool inky darkness that wraps itself round me, a half-moon in the early hours, the first streaks of dawn light, then a crimson blush and dawn chorus, heralding another blue-sky day – and oh the warmth of the golden sun. The hours may have passed, in truth, in an agony of coldness (I'd forgotten my sleeping mat) but they have given me a glimpse of the sacred too. My elation is the by-product of intimate contact with that realm we call night. No separation.

From Ash Dome, via Cae Mabon, the art trail has led me here to this moment. I can feel Gaia's sweet embrace. I'm not unique. I just love the land, and the love is being returned.

# The Secret Place of the Wild Strawberries – Part II

It's midsummer, one year on, and I'm back on Lindisfarne. Some places just have a grip on you. I used to naturally gravitate towards the south: south felt familiar, south felt warm, south felt welcoming. South India is where my ancestors are from. South is in my blood. But I feel that embrace here in this corner of the north now.

This time I'm in a rented cottage on the edge of the village, with a wrap-around garden and a row of towering sycamores along the fence. 'Cottage' is a bit of an understatement; it is a posh, two-storey affair with skylight windows and views across the fields. The sunshine is dazzling but the wind is up, all raw and fierce and pushback. The familiarity is sweet, though.

I've come with my friend Mandy, but things are not quite working out between us. All I want is for her to connect with the insane, wild beauty of the island in the way that I do, to feel as exhilarated by it as I do. This is a healing place, and I want her to be healed. I want to wash away her illness, the cancer she is treating. It is so much easier to focus on the vulnerabilities of others, at the expense of our own. I mean, who is the one who really needs healing? Who *doesn't* need healing? But love has its place to play too. For me the love is split two ways: love for a friend and love for this sacred place. I want to offer the gift of one as a gift to another.

But my offering comes with an unspoken condition: to not enter into the spirit of Lindisfarne is, to my mind, unforgiveable, a sacrilege. I imagine the land – lush at its peak now, every tree in full leaf, every blossom and flower flourishing – would never be as demanding with me. Obviously, wanting a clone of myself for company is asking for a lot. But this past year has changed me and distanced me from my friend. Yet the failure is mine in not doing more to bridge the gap. And in not letting her be herself: the very thing I criticise in those who dismiss me.

Still, there is so much to take in outside: the sparrows and swifts darting about in the garden, and birds that a book I find in the house tells me are pied wagtails. There is another with zebra-like tail feathers that I've not seen before. The birds are tame, and chirp from dawn till dusk, which is a kind of heaven in itself. I'm not sure you can be human and immune to birdsong.

Across the island the wildflowers are out, another kind of ecstasy. That is what it feels like: little pleasure-givers all over the island. On a sloping bank along the walkway to the castle is an entire carpet of valerian, an intense shade somewhere between fuchsia and carmine. They seem happy, their faces to the sea and the sun. They remind me of a choir, singing a silent hymn.

That afternoon we sit in front of the labyrinth with our backs to a grassy slope to protect us against the ferocious wind and lap up the sunshine and gaze at the sea. 'Isn't it amazing? Isn't it perfect?' I say to Mandy. 'Can you feel the magic?'

I demand a display of emotion and I get none. My friend nods quietly. Maybe I mistake her quietness for a lack of interest. Maybe her appreciation comes in a quieter register. In a less hungry, less mystical-seeking one.

'How about a coffee in the village?' she says after a while, shivering. It pains me to be torn away from this spot and I am annoyed but I don't show it – yet. I don't bother to consider that maybe this is as much as she can manage this afternoon, that maybe she is tired from the journey and from mothering a small child.

We bump into Mary Gunn, the guide, on the way back. She greets me by name and I'm flattered by the familiarity.

'Have you looked for the wild strawberries yet?' she asks.

'Not yet, but I plan to,' I say, smiling as a gust of wind buffets us.

'Well, they've been seen upcountry, so you might get lucky.'

I hope so. Just a few days earlier, while reading a novel I'd stumbled upon a Swedish phrase: *Smultronstället*. It translates as 'the Secret Place of the Wild Strawberries'. Metaphorically, it's a healing place, a place you can call home. Such a place doesn't have to be a literal geographical place but a state of mind, although it could also be a place. I love the sound of *Smultronstället*. Maybe that is what I am looking for, I think, promising myself that if there is a single wild strawberry on this island I will find it too. I tell Mandy this story excitedly and she nods. 'Sounds great,' she says, indulging me. 'You have a mission now.' Why isn't she feeling the way I am? My frustration grows.

That evening the sky has a luminous, timeless quality. Through the back windows of the house, while Mandy is resting in her room, I watch a group of children playing in fields, raucous and innocent and carefree as lambs – not a thing you can often say about children these days. I see a flock of tiny birds flying in formation, round and round, swooping and rising as one, like a mini-murmuration.

That night I sleep badly and in the morning the pair of us hover over the kettle in the kitchen awkwardly. There's a tension between us and we decide to go our own ways across the island. The weather is overcast, rain threatening, but I make my way to the North Shore, proud that I have remembered the shortcut Mary Gunn showed me the last time. I can't resist stopping every few minutes to gawp at the wildflowers as I make my way through the dunes. A hare shoots out from nowhere and high-tails it into the marram grass, and then another sprints into a burrow.

I cross the dune valley, and spot Mandy on a boulder, checking her phone for a signal. We wave and I walk on. Phone. Ugh, I think. But then I don't have a six-year-old boy I might be missing. The North Shore looks bleaker and wilder today. The wind is out hunting, with no concessions made for walkers who get in its way today. I notice traces of horse hooves in the sand. In the salt marshes behind the Snook, I search for the wild strawberries. This was the spot.

I search hard, and the world beneath my feet comes alive – strange flora, Lilliputian plants and weeds and wild growths. I am plant illiterate, though. I know nothing of the life here, can't map it or make heads or tails of it. I am alone but I can hear the ghosts of the naturalists hissing in my ear: *If you were indigenous to this island you'd know the names of these weeds, these plants. You're not salt of the earth*. My inability to find a single strawberry is further proof of my inadequacy.

Finally, still strawberry-less, I slink back along the beach till I get to Coves Haven. It's the same as I remembered it: rust-red cliffs to one side, jutting rocks on which seals frolicked on the other. The sky is still iffy, the sea blue and grey, and I begin to come back to myself. It's so still and quiet here, with the receding tide leaving dark flotsam and mussels and seaweed. It's cathedral still, holy still. I listen hard.

I think I can listen better if I lie down in the sand, so I do that. I use my rucksack as a pillow, pull my hood up, close my eyes and yield to the peace. I drift off to sleep. It's the sweetest sleep. When I open my eyes again, maybe fifteen minutes later, I see a couple. Dressed in earthy colours as if to blend in with their surroundings, they are foraging. The pair utter not a word to each other and are respectful of the silence here. Is that a pair of seals I see in the distance? I can't hear their mournful wailing this morning. But I feel serenity on this beach, just as I had on the White Strand of the Monks in Iona. There is so much generosity here in the stillness.

A spell has been cast, though who or what the spellcaster is, I can't say.

When I get up to leave I feel as though my DNA has been altered. I'm as well as I ever have been in my entire life, as sharp and clear-headed and rested and full as it is humanly possible to feel. You move through loneliness and disconnection and you get to this, if you're lucky.

⋆ ⋆ ⋆

I'd planned to walk the labyrinth on the morning of the solstice to bring closure to my year, but the mist and the rain have put an end to that. Barely speaking now, Mandy and I have set off round the island in different directions. I hope we don't meet in the middle. But sod's law, we do. We collide on the footpath near the pyramid at Emmanuel Head. Mandy has been for a swim in the cove and is rejuvenated, exhilarated. Me, I'm soaked and mad at her for not getting that there is more to this island than meets the eye, for not caring about the strawberries, but mad too because she is ill and I don't see her often enough.

Goats appear around us, spectre-like in the mist and watch on, bemused, as we have a huge, storming row. In hindsight our eruption is comical. We're literally the only humans out here and we're spewing like Vesuvius. We will laugh about it three months later, when we have made up. But right now, furious, my friend who right now is not a friend decides to leave the island, and I walk slowly, very slowly, back to the village.

I take a shortcut through a farmer's field and find myself in a pretty walled garden owned by the National Trust, planted to the designs of Gertrude Jekyll. It is a 'summer flowering garden', ridiculously neat and pretty, and it faces the castle. For now it is a mixture of drenched, not-quite-ripe vegetables and fruit, and soggy flowers. Amid the foxgloves and radishes, my eyes are drawn to a lone ripe strawberry among the plants. *A strawberry!* Just the one. It may not be a *wild* strawberry but the island has done its best: it is a strawberry in a wild place. I had nurtured the crazy, fanciful idea that if I were to find just one strawberry, it would be a sign that I was all right, that I was heard, that I had found my true home in that liminal space where all things are possible, in my good, safe Narnia.

The second I have the thought, it leads to another, so big it sends me reeling. I have spent the entire year in pursuit of a deeper, sweeter relationship with this land – feeding my imagination, my hunger for mystical experience, for beauty, for communion, for a connection with the Divine – but something else has been going on too. Until this journey, I had put my truest self into self-imposed exile, cowed by culture and society and feelings of not belonging because of the colour of my skin and my own anxieties. Unconsciously, I'd set off driven by a need to retrieve those lost, mute and unloved parts of myself as much as anything else. To bring them out into the light of day and let them breathe. Every step has been an opportunity to make friends with those parts of myself. The me who is not steeped in the traditions of this country, the me who is an outlier, who wants to be free to love and claim belonging to this land, just as I am. In my quirky way, in a small way, I've sought to bring the Other, both wild and human, in from the cold. In wanting to ritualise my journey, cast a spell, invoke enchantment, I've roused myself from a slumber. Now, here I am, awake, and there is more of me. It is a good feeling. Maybe it's the feeling that comes when the scales are tipped and with enough life experience under your belt, being vulnerable and showing your true colours doesn't scare you as much, even if you are naturally reserved.

In the afternoon, I pace around a giant, empty cottage – I feel Mandy's absence keenly – and finally, once the causeway has closed for the day and I can no longer hear the voices of the day trippers, the sun decides to show herself. I decide to take an evening walk. I head through the churchyard and down a footpath, past two horses in a field, too busy gorging on tall grass to lift their heads. I come to the little strip of beach overlooking St Cuthbert's island, now separated from the island by the tide.

This is how I will honour the solstice, I decide. I sit on a bench and watch the sun's rays burst out from behind a cloud, creating a halo. I wish I hadn't judged my friend, I wish I had been more tolerant. We don't listen to each other enough in this modern world and we don't listen to the murmurings of the land. Every tree, every leaf, every ray of sunlight, every bird, every creature is reaching out to us in their language. *'Please listen…'*

Sitting there on that bench, I make a promise to listen harder. And I am going to give voice to the longings I've guarded within me for a long, long time, from the time I was a little girl drawn to the things I could not see. I am going to begin by writing down everything that has happened to me this year.

# Acknowledgements

A heartfelt thank you to my publisher Jim Martin for believing in me – you gave me the freedom to write the book that I wanted to and that has meant everything to me.

I'm grateful to the whole team at Bloomsbury and in particular my editor Alice Ward, for her patience, enthusiasm and kindness.

My gratitude goes to all who've been a part of my journey – I'm indebted to you for your time, your enthusiasm and encouragement, your insights and your company.

For all those who've buoyed up my morale when it has flagged, thank you and I hope to return the favour one day.

A very special thank you to my local library, Wimbledon, for providing so much inspiration on your shelves and for being such a wonderful place to write quietly.

I am forever grateful to my mother, Sandra, for her love and support. And to my late father for telling me I could be anything I wanted to be.

To those reading, thank you so much for taking the time.

# Index